Seismicity and Seismic Risk in the Offshore North Sea Area

NATO ADVANCED STUDY INSTITUTES SERIES

Proceedings of the Advanced Study Institute Programme, which aims
at the dissemination of advanced knowledge and
the formation of contacts among scientists from different countries

The series is published by an international board of publishers in conjunction
with NATO Scientific Affairs Division

A	Life Sciences	Plenum Publishing Corporation
B	Physics	London and New York
C	Mathematical	D. Reidel Publishing Company
	and Physical	Dordrecht, Boston and London
	Sciences	
D	Behavioural and	
	Social Sciences	Martinus Nijhoff Publishers
E	Engineering and	The Hague, London and Boston
	Materials Sciences	
F	Computer and	Springer Verlag
	Systems Sciences	Heidelberg
G	Ecological Sciences	

Series C – Mathematical and Physical Sciences

Volume 99 – Seismicity and Seismic Risk in the Offshore North Sea Area

Seismicity and Seismic Risk in the Offshore North Sea Area

Proceedings of the NATO Advanced Research Workshop,
held at Utrecht, The Netherlands, June 1-4, 1982

edited by

A. REINIER RITSEMA
Royal Netherlands Meteorological Institute (KNMI), De Bilt, The Netherlands

and

AYBARS GÜRPINAR
Specialized Engineering Consulting Service (SPECS), Brussels, Belgium

D. Reidel Publishing Company

Dordrecht : Holland / Boston : U.S.A. / London : England

Published in cooperation with NATO Scientific Affairs Division

Library of Congress Cataloging in Publication Data

Main entry under title:

Seismicity and seismic risk in the offshore North Sea area.

    (NATO advanced study institutes series. Series C, Mathematical and
physical sciences ; v. 99)
        "Workshop on the Seismicity and Seismic Risk in the Offshore North Sea
        Area"—introd.
        "Published in cooperation with NATO Scientific Affairs Division."
        Includes indexes.
        1. Seismology—North Sea—Congresses. 2. Earthquake prediction—
Congresses. I. Ritsema, A. Reinier (Anne Reinier) II. Gürpinar, Aybars.
III. Workshop on the Seismicity and Seismic Risk in the Offshore North Sea
Area (1982 : Utrecht, Netherlands) IV. North Atlantic Treaty Organization.
Scientific Affairs Division. V. Series.
QE536.2.N67S44   1983   551.2'2'0916336   82-21624
ISBN 90-277-1529-7

Published by D. Reidel Publishing Company
P.O. Box 17, 3300 AA Dordrecht, Holland

Sold and distributed in the U.S.A. and Canada
by Kluwer Boston Inc.,
190 Old Derby Street, Hingham, MA 02043, U.S.A.

In all other countries, sold and distributed
by Kluwer Academic Publishers Group,
P.O. Box 322, 3300 AH Dordrecht, Holland

D. Reidel Publishing Company is a member of the Kluwer Group

SEISMICITY AND SEISMIC RISK IN THE OFFSHORE NORTH SEA AREA

ADVANCED RESEARCH WORKSHOP

June 1–4, 1982, Hotel Des Pays-Bas, Utrecht, The Netherlands

Sponsored by and organized on behalf of the

SCIENTIFIC AFFAIRS DIVISION
NORTH ATLANTIC TREATY ORGANIZATION

Co-sponsored by:

Royal Netherlands Meteorological Institute (KNMI)

Scientific Directors:

A. R. Ritsema
KNMI, De Bilt
Netherlands

A. Gürpınar
SPECS, Brussels
Belgium

Advisory Committee:

P. W. Burton
IGS, Edinburgh
United Kingdom

J.-M. van Gils
ORB, Brussels
Belgium

P. Selnes
NGI, Oslo
Norway

Invited Lecturers:

P. A. Ziegler
SIPM, 's-Gravenhage
Netherlands

F. Ringdal
NORSAR, Kjeller
Norway

P. Melchior
ORB, Brussels
Belgium

C. W. A. Browitt
IGS, Edinburgh
United Kingdom

H. L. Koning
SML, Delft
Netherlands

N. N. Ambraseys
Imp. Coll., London
United Kingdom

TABLE OF CONTENTS

Section 2: Seismicity

Section 3: Tides, Ocean Waves and Sea Level Changes

Section 4: Instrumentation

Section 5: Soilmechanics, Liquefaction, Geotechnology

Section 6: Risk Analysis

INTRODUCTION

The Workshop on the Seismicity and Seismic Risk in the Offshore North Sea Area was intended to bring together experts from
a variety of disciplines as well as interest groups with involve-
ment in siting, design and construction of offshore structures
in the region. Participants came from the fields of geology,
seismology, oceanography, geotechnical and structural engineering
and risk analysis. The wide range of participant affiliations
included institutes, observatories, universities, oil companies,
consultants and insurance firms. All nationalities around the
North Sea were present, in addition to some experts from outside
the region. All participants were present on the basis of personal
invitation.

The idea of organizing the Workshop stemmed from considerations, such as:
- the rapidly increasing material and personel investments and
  versatility of type of structures in the basin during the past
  decade;
- the present-day important role of the North Sea oil and gas
  production in the economy of Western Europe; and
- the increase of potential environmental risks in the region.

Although devastating earthquakes are almost unknown in the
area and seismic hazard is not great, the seismic risk grows with
the growing size and number of structures in the area. The study
of the potential seismic risks, therefore, cannot be neglected
any more. The siting and design of offshore platforms and submarine
pipelines are controlled by the degree of their vulnerability as
well as the seismic hazard in the region.

The scientific challenge for the Workshop came from the fact
that the North Sea basin area exhibits certain unique and compli-
cating characteristics related to its seismotectonic setting.
Among these are the following:
- the large offshore area with few significant historical events,
  seemingly not exhibiting a definite pattern;
- the intra-plate character of the region, i.e. absence of plate
  boundaries which clearly identify causative structures;
- the relatively short time of sensitive instrumental observations;
- the large uncertainties associated with the seismic parameters
  such as frequency-magnitude relationships and maximum credible
  earthquake;
- the very slow attenuation of seismic effects which tends to
  increase the seismic hazard in general and spread it more
  uniformly throughout the area; and

A. R. Ritsema and A. Gürpınar (eds.), Seismicity and Seismic Risk in the Offshore North Sea Area, xi–xii.
Copyright © 1983 by D. Reidel Publishing Company.

- the little or incomplete isoseismal information, due to the
  offshore nature of the region.

These characteristics and the resulting uncertainties in the
seismic parameters tend to increase structural design values
significantly for low levels of prescribed probabilities of
exceedance. The Workshop, in contradiction to what is often the
case in the field of seismicity, was not held subsequent to a
great earthquake in the area. This proved useful for the discussions
to remain relatively free of bias and distortion.

The program of the Workshop comprised six main topics, each
introduced by a keynote speaker. These were followed by discussions
consisting of short presentations and comments on the introductory
lecture by members of the Panel in question. Sessions were then
concluded by free discussion from the floor. This program of the
Workshop is reflected in the structure of the present proceedings.
Discussions are not reproduced verbally; several questions and
answers were assimilated in the final manuscripts provided by the
speakers. The gist of the discussions following the latest section
on Seismic Risk Assessment is encorporated in the section on
Conclusions and Recommendations.

ACKNOWLEDGEMENTS

We are grateful to the NATO Scientific Affairs Division
which provided a grant to finance the Workshop.
The excursion of the group to the Royal/Shell Exploration and
Production Laboratory (KSEPL) in Rijswijk, and the reception by
the Director-in-Chief of the Royal Netherlands Meteorological
Institute (KNMI) at De Bilt are gratefully acknowledged. Important
logistic support during and after the Sessions was given by
G. Houtgast, J. van Gend and Mrs. J. van Bodegraven, all of KNMI.
Editorial help in the compilation of the present proceedings was
obtained from G. Houtgast and Mrs. J. van Bodegraven.

De Bilt/Brussels, July 1982

A. Reinier Ritsema
Aybars Gürpınar

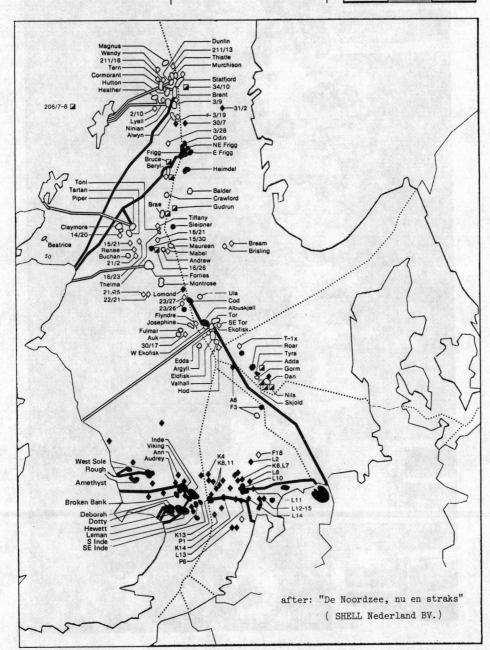

OIL AND GAS IN THE NORTH SEA BASIN 1980 SITUATION

Oil

Crude-oilpipeline

⬡ Oilfield
◇ Oilfind
◪ Oil and gas find

Gas

Gaspipeline

● Gasfield
◆ Gasfind
● Condense find

National boundaries
on continental shelf

km
0   100   200   300

Magnus
Wendy
211/16
Tern
Cormorant
Hutton
Heather

206/7-8

2/10
Lyell
Ninian
Alwyn

Frigg
Bruce
Beryl

Toni
Tartan
Piper

Claymore
14/20

Beatrice
50

15/21
Renee
Buchan
21/2

16/23
Thelma

21/25
22/21

Dunlin
211/13
Thistle
Murchison
Statfjord
34/10
Brent
3/9   31/2
3/19
30/7
3/28
Odin
NE Frigg
E Frigg

Heimdal

Balder
Crawford
Gudrun

Brae

Tiffany
Sleipner
16/21
15/30
Maureen
Mabel
Andrew
16/26
Forties
Montrose

Lomond   Ula
23/27   Cod
23/26   Albuskjell
Flyndre   Tor
Josephine   SE Tor
Fulmar   Ekofisk
Auk
30/17
W Ekofisk

Edda
Argyll
Eldfisk
Valhall
Hod

A6
F3

Bream
Brisling

T-1x
Roar
Tyra
Adda
Gorm
Dan

Nils
Skjold

Inde
Viking
Ann
Audrey

West Sole
Rough

Amethyst

Broken Bank

Deborah
Dotty
Hewett
Leman
S Inde
SE Inde

K4
K8,11

K13
P1
K14
L13
P6

F18
L2
K6,L7
L8
L10

L11
L12-15
L14

after: "De Noordzee, nu en straks"
(SHELL Nederland BV.)

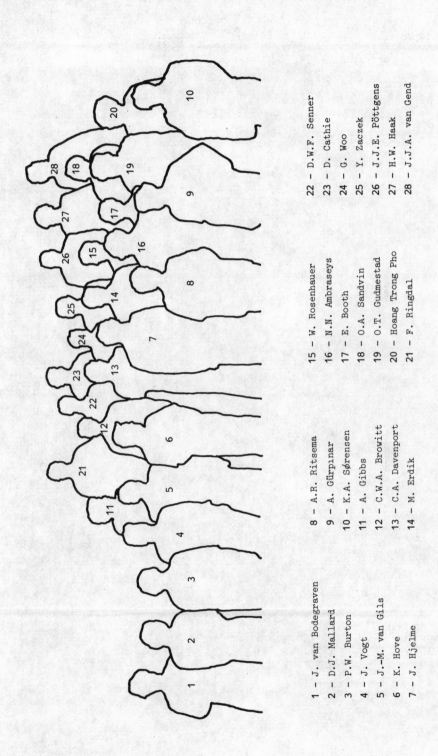

1 – J. van Bodegraven
2 – D.J. Mallard
3 – P.W. Burton
4 – J. Vogt
5 – J.–M. van Gils
6 – K. Hove
7 – J. Hjelme

8 – A.R. Ritsema
9 – A. Gürpinar
10 – K.A. Sørensen
11 – A. Gibbs
12 – C.W.A. Browitt
13 – C.A. Davenport
14 – M. Erdik

15 – W. Rosenhauer
16 – N.N. Ambraseys
17 – E. Booth
18 – O.A. Sandvin
19 – O.T. Gudmestad
20 – Hoang Trong Pho
21 – F. Ringdal

22 – D.W.F. Senner
23 – D. Cathie
24 – G. Woo
25 – Y. Zaczek
26 – J.J.E. Pöttgens
27 – H.W. Haak
28 – J.J.A. van Gend

List of Participants

ADRICHEM BOOGAERT, H.A. v., Rijks Geologische Dienst, Spaarne 17, 2011 CD Haarlem, The Netherlands; tel.: 23-319362.

AHORNER, L., Erdbebenstation der Universität Köln, Vinzenz Pallotti Strasse 26, D-506 Bensberg, F.R.G.; tel.: 2204-81343.

AMBRASEYS, N.N., Dept. of Civil Engineering, Imperial College of Science, London, SW7, 2BU, England; tel.: 1-5895111 ext. 1377.

BOOTH, E., Ove Arup & Partners, Consulting Engineers, 13 Fitzroy Street, London, W1P 6BQ, England; tel.: 1-7349321; telex: 263935.

BOUWS, E., K.N.M.I., Division of Oceanographic Research, P.O. Box 201, 3730 AE De Bilt, The Netherlands; tel.: 30-766911; telex: 47096 knmi nl.

BROWITT, C.W.A., Institute of Geological Sciences, Murchison House, West Mains Road, Edinburgh EH9 3LA, Scotland, United Kingdom; tel.: 31-6771000.

BURTON, P.W., Institute of Geological Sciences, Murchison House, West Mains Road, Edinburgh EH9 3LA, Scotland, United Kingdom; tel.: 31-6771000.

CATHIE, D., D'Appolonia Consulting Engineers Inc., Blv. du Souverein, Brussels, Belgium; tel.: 2-6604990.

CSIKÓS, I., K.N.M.I., Division of Geophysics, P.O. Box 201, 3730 AE De Bilt, The Netherlands; tel.: 30-766911; telex: 47096 knmi nl.

DAVENPORT, C.A., University of Strathclyde, Dept. of Applied Geology, James Weir Building, Montrose Street, Glasgow G1 1XJ, Scotland, United Kingdom; tel.: 41-5524400.

ERDIK, M., Director Earthquake Engineering Research Institute, Middle East Technical University, Ankara, Turkey; tel.: (90-41) 237100; telex: 42761 odtk tr.

EWOLDSEN, H.M., Woodward & Clyde, 35 Markham Square, London SW3, England; tel.: 1-5810285.

GIBBS, A., British National Oil Corporation, Exploration Division, 150 Vincent Street, Glasgow G2 5LJ, Scotland, United Kingdom; tel.: 41-2042566; telex: 776268.

GIESEN, M.H., Rijks Geologische Dienst, Spaarne 17, 2011 CD Haarlem,
    The Netherlands; tel.: 23-319362.

GIJT, J.G. de, Fugro bv, Geotechnical Engineers and Surveyors,
    P.O. Box 63, 2260 AB Leidschendam, The Netherlands; tel.:
    70-209250; telex: 31010.

GUDMESTAD, O.T., a/s Norske Shell, P.O. Box 40, N-4056 Tananger,
    Norway; tel.: 4-696488; telex: 33368 shelp n.

GÜRPINAR, A., SPECS, Place Stéphanie 10, 1050 Brussels, Belgium;
    tel.: 2-512226; telex: 61344 specs b.

HAGEMAN, B.P., Director Rijks Geologische Dienst, Spaarne 17,
    2011 CD Haarlem, The Netherlands; tel.: 23-319362.

HEALY, P., D'Appolonia Consulting Engineers Inc., Blv. du Souverein,
    Brussels, Belgium; tel.: 2-6604990.

HJELME, J., Geodaetisk Institut, Seismisk Afdeling, Gamlehave Alle
    22, DK-2920 Charlottenlund, Denmark; tel.: 1-631833; telex:
    15184 seismo dk.

HOANG Trong Pho, Centre Seismologique Européo-Med., 5 Rue René
    Descartes, F-67084 Strasbourg Cedex, France; tel.: 88-614820;
    telex: 890826 csem f.

HOUTGAST, G., K.N.M.I., Division of Geophysics, P.O. Box 201,
    3730 AE De Bilt, The Netherlands; tel.: 30-766911; telex:
    47096 knmi nl.

HOVE, K., Det Norske Veritas, P.O. Box 300, 1322 Høvik, Norway;
    tel.: 2-129900.

KONING, H.L., Laboratorium voor Grondmechanica, P.O. Box 69,
    2600 AB Delft, The Netherlands; tel.: 15-569223.

KRUISMAN, G., Rutten & Kruisman Consulting Engineers bv, Geest-
    brugweg 146, 2281 CM Rijswijk, The Netherlands; tel.:
    70-906890.

MALLARD, D.J., Civil Engin. Branch, Generation Development and
    Construction Division (CEGB), Barnwood, Gloucester GL4 7RS,
    United Kingdom; tel.: 452.652967; telex 43501 cegbgd g.

MELCHIOR, P., Observatoire Royal, 3 Avenue Circulaire, 1180-
    Brussels, Belgium; tel.: 2-3943802; telex: 21565 obsbel b.

NOLET, A.M.H., Instituut voor Aardwetenschappen, Vening Meinesz
    Laboratorium, Budapestlaan 4, 3584 CD Utrecht, The Netherlands;
    tel.: 30-535088.

PÖTTGENS, J.J.E., Staatstoezicht o/d Mijnen, Apollolaan 9,
    6411 BA Heerlen, The Netherlands; tel.: 45-718188.

RINGDAL, F., NORSAR, P.O. Box 51, 2007 Kjeller, Norway; tel.:
    2-716915; telex: 18147 kcin n.

RITSEMA, A.R., K.N.M.I., Chief Division of Geophysics, P.O. Box
    201, 3730 AE De Bilt, The Netherlands; 30-766911; telex:
    47096 knmi nl.

RONDE J.G. de, Directie Waterhuishouding en Waterbeweging, P.O.
    Box 20907, 2500 EX Den Haag, The Netherlands; tel.: 70-
    889370.

ROSENHAUER, W., Internationale Atomreaktorbau GmbH (INTERATOM),
    Friedrich Ebert Str., D-5060 Bergisch Gladbach 1, F.R.G.;
    tel.: 2204-48928; telex: 8878457 iagl d.

SANDVIN, O.A., NTNF/NORSAR, P.O. Box 51, 2007 Kjeller, Norway;
    tel.: 2-716915; telex 18147 kcin n.

SCHITTEKAT, M., TRACTIONEL, Dept. Engin., Service Mécanique des
    Sols, Rue de la Science 31, 1040 Brussels, Belgium; tel.:
    2-2344111.

SELNES, P., Norges Geotekniske Institutt, P.O. Box 40, Tåsen,
    Oslo 8, Norway; tel.: 2-230388; telex: 19787 ngi n.

SENNER, D.W.F., McClelland Engineers. s.a., McClelland House, Chantry
    Place, off Headstone Lane, Harrow, Middlesex, HA3 6NY, United
    Kingdom; tel.: 1-4212411; telex: 925759 mesa g.

SØRENSEN, K.A., Cowiconsult, Teknikerbyen 45, DK-2880 Virum,
    Denmark; tel.: 2-857311; telex: 37280 cowi dk.

VAN GILS, J.M., Observatoire Royal, 3 Avenue Circulaire,
    1180 Brussels, Belgium; tel.: 2-3943802; telex: 21565 obsbel b.

VLAAR, N.J., Instituut voor Aardwetenschappen, Vening Meinesz
    Laboratorium, Budapestlaan 4, 3584 CD Utrecht; tel.:
    30-535085.

VOGT, J., BRGM, 204 Route de Schirmeck, F-6720 Strasbourg, France;
    tel.: 88-301262; telex: 880322 brgm f.

WOO, G., Principia Mecanica Ltd., Newton House, 50 Vineyard Path,
    East Sheen, London SW14 8ET, England; 1-8787933; telex:
    894694 prince g.

WOOD, R.M., Principia Mecanica Ltd., Newton House, 50 Vineyard
    Path, East Sheen, London SW14 8ET, England; tel.: 1-8787933;
    telex: 894694 prince g.

ZACZEK, Y., Comp. Gen. d'Entreprises Electriq. et Industr.,
    1 Place du Trône, 1000 Brussels, Belgium; tel. 2-5117240;
    telex: 25358 ebl b.

ZIEGLER, P.A., Shell Internationale Petroleum Maatschappij (SIPM)
    EP/11, Postbus 162, 2501 AN Den Haag, The Netherlands;
    tel.: 70-771756.

OBSERVERS:

    J.A. As
    Th. de Crook   }   K.N.M.I., De Bilt.
    H.W. Haak

    A.P. van den Berg
    S. Cloetingh   }   V.M.L. , Utrecht.
    B. Dost
    R. Wortel

NATO ADVANCED RESEARCH WORKSHOP
ON
THE SEISMICITY AND SEISMIC RISK IN THE OFFSHORE NORTH SEA AREA

1-4 June, 1982
Utrecht, The Netherlands

CONCLUSIONS AND RECOMMENDATIONS

As a result of the discussions the following conclusions and
recommendations for further work in the field were put forward.
Their subsequent formulation is under responsability of the editors.
More detailed and explicit suggestions can be found in a number
of individual contributions to the Proceedings of the Workshop.

The interdisciplinary approach of the meeting aiming at the
elimination of information lacuna, the testing of ideas and the
bridging of communication gaps has proved to be fruitful. An inten-
sified exchange of information between scientific and application
oriented private organizations is of great importance and will be
of great help in the future.

The main conditions for any comprehensive study in the field
are the completeness and the general availability of basic data
to be used.

Recognizing that Seismic Risk depends on Seismic Hazard,
Vulnerability and Economic Value of the structure, as expressed
by Ambraseys as follows:

RISK = HAZARD x VULNERABILITY x VALUE ,

the conclusions and recommendations have been listed in such a
way to best complement the existing information on these topics
of seismic hazard assessment and vulnerability studies. The
following areas of special interest and further study are identified:

1. A certain information gap between geologists, seismologists
   and design engineers was observed. It was also noted that
there is no uniform approach in statistical methodologies, such
as used in the determination of the maximum credible earthquake,

xxi

A. R. Ritsema and A. Gürpinar (eds.), Seismicity and Seismic Risk in the Offshore North Sea Area, xxi–xxiv.
Copyright © 1983 by D. Reidel Publishing Company.

other earthquake thresholds, and questions of accumulated risk in
the case of simultaneous loading by earthquakes, tides and ocean
waves. Suggestions for improvement included the following points:

- Exchange of knowledge between active workers in the fields of
  geology, seismology, geotechnology and engineering should be
  further developed.
- Co-ordination within and in between national and international
  seismological services, institutes, universities, research
  centres and private industry should be encouraged.
- Basic and mid-career training of engineers and other design
  professionals in seismic safety should be undertaken.
- The data exchange for relevant events should be formalized and
  made more rapid via telex and view data, including the transmission of complete wave forms.

2.   The need for a comprehensive data-bank for relevant informa-
      tion of all kinds was intensely felt during the Workshop.
Such a bank of reliable, up-to-date and readily available data
should be established applying to all aspects of hazard assessment:

- Tectonic data, more specifically the complete map of recent and
  quaternary movements and locations of active faults.
  Tectonic studies, especially of the Quaternary period should be
  carried out for the basin and adjacent regions, including careful mapping of the recently active faults associated with the
  regional stressfield, saltdomes, gravitational instability
  structures or others.
  The question of the feasability to compile all relevant information from governmental as well as private sources in a comprehensive way, should be investigated. In any case, for specific
  events or areas detailed information should be sought and compiled from appropriate authorities.
- The seismic data list, comprising all historical events in the
  region unified in magnitude; and the list of seismic stations,
  their instrumentation and detection and location capabilities.
  The data from all North Sea countries should be brought together
  into comprehensive catalogues.
- A soil-profile inventory, which is urgently needed in the field
  of geotechnology.
  The question of the possible representative response spectra
  and time-histories of typical North Sea events should subsequently
  be assessed.
- Comprehensive case histories should be made available as a tool
  for interpretation of future events.

3.   Important suggestions were made in <u>instrumentation</u>, where
      the calibration of existing systems and the increase of
detection and location capabilities are stressed to be accomplished
by an extension of on- and offshore networks.
The need of nearfield data for depth determinations is expressed,
as are the increase of dynamic range of strong motion instruments
to longer periods in the range of that of the very long period
structures already existing and envisaged for the region. Sea-
bottom seismometers possibly could be made 'multi-purpose' for
the monitoring of platform movements and other disturbances in
the local environment.

- A submarine seismometer as standard equipment with each existing
  and future field in the region should greatly help the identification of small earthquakes and their depth in the region.
  The installation and maintenance of such sensors by the responsible platform authorities is strongly recommended.
- The use of surface wave analysis for the discrimination of
  depth of focus is recognized and should be further developed.

4.   For a certain number of parameters in the field of geotechno-
      logy there is a need for <u>active research</u>, such as:

- A special study should be directed to provide materials for the
  determination of attenuation parameters in the area.
- Laboratory and field experiments should be carried out for an
  improved understanding of the behaviour of foundation materials
  under earthquake loading. This also includes the development and
  testing of laboratory and field instrumentation to monitor the
  behaviour of foundation materials under seismic loading, vibrational as well as displacement controlled. It should be noted
  in this context that Northwestern European countries historically have an excellent record in the field of static structural and soil mechanics experiments, and that very little
  effort would be required to extend this into the dynamics field.
- Vulnerability studies should be performed for the existing
  offshore platforms and submarine pipelines; both laboratory
  experimentation and field monitoring may be used for this
  purpose.

5.   A final suggestion concerns the further evaluation of present
      and future work in the field:

- A <u>second Workshop</u> on the same topic should be held in about two
  years time to assess the developments made in the matter. The
  need to undertake organised interdisciplinary and multinational
  research was emphasized and the procurement of funds from appropriate parties including NATO and private industry (oil companies)

to support this effort, was expressed as a prerequisite.

These Recommendations should be brought to the attention of
interested individuals and appropriate authorities and agencies
for timely action.

SEISMIC ACTIVITY 1900-1980 IN THE NORTH SEA BASIN

(data from NEIS, EMCS and local sources)

LATITUDE : 50.000N - 66.000N   MAGNITUDE: MB MS ML M I ALL
LONGITUDE :   5.000W - 12.000E   DEPTH(KM): ALL
REMARKS   : NORTH SEA AREA   PERIOD   : 1900 - 1980
                                      F-E   :

TOTAL NUMBER OF EARTHQUAKES FOR THIS SELECTION : 196

INTENSITY

9 - 10	
7 - 8	
5 - 6	
3 - 4	
1 - 2	

MAGNITUDE

7 - 7.9	
6 - 6.9	
5 - 5.9	
4 - 4.9	
3 - 3.9	
2 - 2.9	
- 1.9	

DEPTH (KM)

0 - 70	71 - 300	301 - 700

TYPE

MB	MS	ML	M	UNKNOWN

MB   Magnitude determined by body-waves.
MS   Magnitude determined by surface-waves.
ML   Local magnitude (distance < 600 km.)
M   Value obtained from various sources,
     unspecified magnitude type
     but generally MS.

K N M I

Section 1

GEOLOGY AND TECTONICS

GEOLOGY AND TECTONICS

Invited Paper

TECTONICS OF THE NORTH SEA BASIN AND THEIR POSSIBLE RELATION TO
HISTORICAL EARTHQUAKES

Peter A. Ziegler
Shell Internationale Petroleum Maatschappij B.V.

Although the North Sea is considered to be a tectonically
stable area, earthquakes, with epicentres located offshore,
have been recorded by seismological stations in the surrounding
onshore areas. It is the objective of this paper to analyse the
possible relation between the tectonics of the North Sea and
its present level of seismic activity.

The North Sea occupies a large part of the intracratonic
North-West European Basin, which extends from the Atlantic
shelves of western Norway and the Orkney Isles, through the
lowlands of Germany and Denmark, to Poland. Sedimentary rocks
contained in this basin range in age from Permian to
Quaternary, and in the North Sea attain a maximum thickness of
8 km. The floor of this basin is formed by Precambrian and
Caledonian metamorphics and intrusives, which are overlain by
Cambro-Silurian, Devonian and Carboniferous sediments,
respectively.
As a result of the intense exploration efforts of the oil
industry, the tectonic and stratigraphic framework of the
sedimentary basins underlying the North Sea and adjacent
onshore areas is well-known. For a summary of the tectonics of
the North Sea, and as a guide to the more specialised
literature, the reader is referred to Illing and Hobson (1981)
and Ziegler (1982) (3,8).

1. SUMMARY OF THE NORTH SEA BASIN EVOLUTION

The tectonic framework of the North Sea area is very complex.
It is the result of a long geological evolution during which

3

A. R. Ritsema and A. Gürpınar (eds.), Seismicity and Seismic Risk in the Offshore North Sea Area, 3–13.
Copyright © 1983 by D. Reidel Publishing Company.

periods of increased tectonic activity alternated repeatedly
with periods of relative tectonic quiescence. Although major
faults involving the basement have been mapped in the North
Sea, these appear to be inactive at present. The complexity of
the North Sea tectonics can be appreciated only when viewed in
the context of its geological evolution.

During its geological evolution, the megatectonic setting of
the North Sea area changed repeatedly. Thus, in time, a number
of genetically different basins developed and were stacked on
top of each other.

Following its Late Silurian cratonisation, the North Sea area
was affected by tensional and wrench tectonics. During the
Devonian, these gave rise, for instance, to the subsidence of
the Midland Valley Rift and the Orcadian Basin, and the
development of the large-scale sinistral Great Glen wrench
fault.
In the northern British Isles, tensional tectonics, related to
early rifting phases in the Norwegian-Greenland Sea, persisted
till the Late Carboniferous. The southern North Sea area, on
the other hand, formed part of the Late Carboniferous Variscan
fore-deep basin. During the latest Westphalian, the Devonian
and Carboniferous rifts of the British Isles became partly
inverted in response to compressional stresses that were
exerted on the foreland during the late phases of the Variscan
orogeny. Stephanian and Early Permian times corresponded to a
period of intense wrench faulting, rifting and volcanism (e.g.
Oslo rift). These movements were induced by a change in the
convergence direction between Gondwana and Laurasia during the
terminal phases of the Hercynian diastrophism.

The Late Permian corresponded to a time of relative tectonic
quiescence during which the Rotliegend and Zechstein series
accumulated in the Variscan foreland. These strata contain
major amounts of halite.
From the earliest Triassic to Early Palaeocene the evolution of
the North Sea area was dominated by rifting processes that were
related to the gradual opening of the Tethys, the North
Atlantic and the Norwegian-Greenland Sea. In the North Sea,
regional crustal extension gave rise to the subsidence of a
complex system of grabens and halfgrabens, of which the Viking
Graben and the Central Graben are the most important (Fig. 1).
This graben system remained active until crustal separation was
achieved in the Norwegian-Greenland Sea during the Late
Palaeocene. The alpine diastrophism is reflected in the North
Sea area by the inversion of, for instance, the southern parts
of the Central Graben and of the Danish Trough.
The Late Palaeocene to recent development of the North Sea area
was characterised by a relatively quiet tectonic regime and the

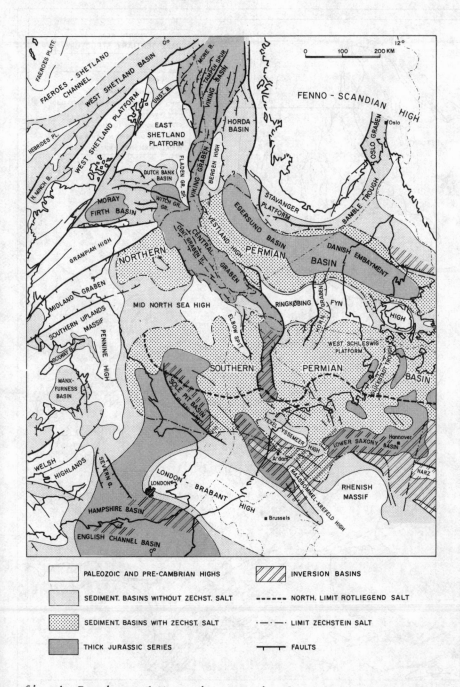

fig. 1. Permian and Mesozoic tectonic elements North Sea area.

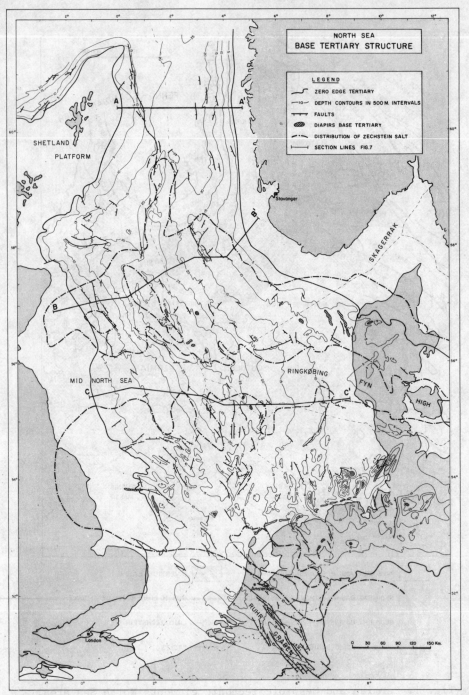

fig. 2

gradual subsidence of the Cenozoic North Sea Basin. In its
central parts, Tertiary and Quaternary strata reach a maximum
thickness of 3500 m.

Each of these tectonic cycles left its mark on the tectonic
framework of the North Sea area. In time, fractures and faults
emplaced during earlier phases of crustal deformation, became
reactivated during subsequent deformation phases.

The following discussion deals with the development of the
Cenozoic North Sea Basin. The processes that governed its
evolution are apparently still going on, and thus may be
responsible for the earthquakes that have occurred in the North
Sea during historical time.

2. CONFIGURATION OF THE CENOZOIC NORTH SEA BASIN

The geometry of the Ceonozoic North Sea Basin is described by
the depth map of the base of the Tertiary sequence of clastics
(Fig. 2). On this map, the zero-edge of the Permian salts is
indicated with a heavy stippled line. This map is based on
extensive reflection seismic surveys that have been calibrated
by hundreds of wells.

The axis of the Cenozoic North Sea Basin coincides closely with
that of the underlying Mesozoic rift system. In cross-section,
the Cenozoic basin is saucer-shaped and almost symmetrical
(fig. 3). Only a very few of the faults that controlled the
Mesozoic rifts also affect the base of the Tertiary clastic
series, and those that do, all die out rapidly in the Paleogene
strata. To the author's knowledge, none of the faults involving
the basement extend upward through the entire Cenozoic series.
This indicates that these faults were not reactivated during
the Neogene and Quaternary.

The last rifting phase in the North Sea is stratigraphically
dated as intra- Paleocene. It caused only a mild reactivation
of the border faults of the Viking Graben and the Central
Graben and also of the Moray Firth fault system.
The Late Paleocene to recent evolution of the North Sea area
was characterised by regional downwarping of the basin floor.
Paleocene, Neogene and Quaternary strata generally increase in
thickness from the margins towards the centre of the basin
(Fig. 4). Notable exceptions, however, are the major Oligocene
deltas that extended from the Shetland platform into the
northern North Sea and the Paleocene and Eocene series, in
which both the turbiditic basinal facies stand out by their
thickness. The North Sea Basin continued to subside to the
present, as illustrated by the Quaternary strata, which locally

8

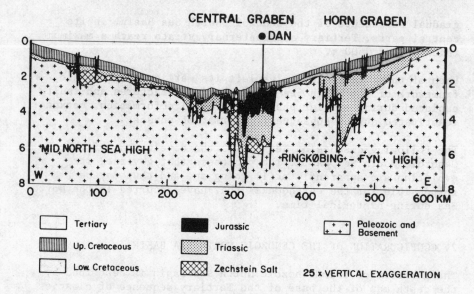

fig. 3. Cross-section through central North Sea latitude 55°
30' N

can reach a thickness of several hundreds of metres (1).

Syndepositional tectonics, related to the gravitational
instability of deltaic complexes that prograded over
undercompacted pro-delta clays, play only a subordinate role in
the North Sea.
In the area of the Permian salt basins, diapirism, and locally
also the solution of the Rotliegend and Zechstein salts,
affected the deposition of the Cenozoic strata by giving rise
to local subsidence anomalies; these interrupt the smooth
pattern of the Cenozoic isopachytes. Salt-induced tectonic
movements are locally still active, causing faults to extend to
the sea bed, and indeed some diapirs have positive expressions
at the sea-floor. Whether such salt movements could give rise
to shallow earthquakes, however, is not known to the author.

3. ISOSTATIC UPLIFT OF FENNOSCANDIA

The margin of the North Sea Basin towards Norway and Sweden is
erosional. Cenozoic and Mesozoic sediments are deeply eroded in
the glacially scoured Norwegian and Skagerrak trenches. The
monoclinal dip and the thickness of the Late Cretaceous and

Cenozoic strata near their erosinal edge, as well as their low
sand content, suggest that their respective shorelines were
located at some distance inland from the present coastlines.
Erosion of these strata reflects the Pleistocene deformation of
Fennoscandia during glaciations as well as its isostatic
updoming in response to the unloading of ice from the crust
during interglacial and postglacial times.

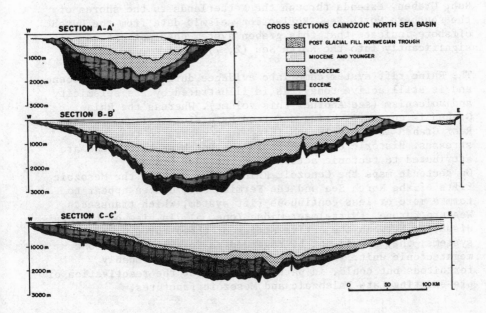

fig. 4. Cross-sections through Cenozoic North Sea basis (for
        location see fig. 2)

The latter process is currently still active and is probably
responsible for the earthquakes which occur in the areas that
are being .uplifted (including the north-eastern margin of the
North Sea).
The strong truncation of the Mesozoic and Cenozoic series along
the margins of the Fennoscandian uplift illustrates that the
post-glacial rebound overcompensated for ice-loading. In other
words, the surface of Fennoscandia is at present uplifted above
the level it had before its glacial loading with ice. The
mechanisms governing this isostatic uplift are still the
subject of debate, but probably involve viscous flow in the
mantle (6), possibly accompanied by phase changes, and elastic
deformation of the crust. This is in keeping with the focal

depth of earthquakes recorded in the onshore areas of Norway and Sweden (see Ringdal, this volume).

The post-glacial uplift of Scotland is less intense and affects the North Sea Basin only marginally.

4. RHINE RIFT SYSTEM

The northwestern branch of the Rhine-Rhône rift system, the Ruhr Graben, extends through the Netherlands to the shores of the southern North Sea. Reflection seismic data from the Dutch offshore indicate that this graben system does not extend significantly into the North Sea (2, 7).

The Rhine rift system came into evidence during the Late Eocene and is still active today, as is illustrated by its seismicity and volcanism (see Ahorner this volume). Whereas the Rhine Graben is currently being deformed by a sinistral shear, the Ruhr Graben is actively subsiding in response to tensional stresses. Historic earthquakes in the Dutch onshore areas are attributed to tectonic activity of the Ruhr Graben.
On tectonic maps the Cenozoic Rhine-Rhône Graben, the Mesozoic rifts of the North Sea and the Permian Oslo Graben appear to form a more or less continuous rift system, which transsects Western Europe. (Mittelmeer-Mjöse Zone, 5). In view of the age disparity between the active stages of these three graben systems, they should not be regarded as forming part of one megatectonic unit. Their apparent alignment is probably fortuitous but could, in part, also reflect the reactivation of pre-existing Late Paleozoic and Mesozoic fractures.

5. SUBSIDENCE MECHANISM OF THE CENOZOIC NORTH SEA BASIN

During its Triassic to Cretaceous active rifting stage, the North Sea Graben system subsided in response to crustal stretching and concomitant, thermally induced "subcrustal erosion". Crustal stretching was accompanied by block faulting, involving the basement with individual faults, having throws of up to several thousands of metres.

With the decrease of rifting activity during the Late Cretaceous, the thermal disturbance induced by regional crustal stretching began to decay. This caused the gradual subsidence of the wider North Sea area. Following the mid-Paleocene termination of all rifting activity in the North Sea, cooling and contraction of the lithosphere, and its isostatic adjustment to water and sediment loading, became the principal

(sole?) mechanism that governed the Cenozoic structural
evolution of the North Sea Basin.

Regional downwarping of the crust apparently did not induce a
reactivation of the faults that controlled the relief of the
Mesozoic grabens.
During the Paleocene and Eocene, the subsidence rates of the
North Sea Basin probably exceeded sedimentation rates, thus
causing its deepening. In the Oligocene, subsidence and
sedimentation rates remained more or less in balance in the
central parts of the North Sea. Mio-Pliocene sedimentation
rates began to accelerate, however, as a result of the
development of the North European river system. Gradual
shallowing of the North Sea basin during the Pliocene was
accompanied by accelerated subsidence. This reflects the
isostatic adjustment of the lithosphere to sedimentary loading.
In the Quaternary, the North Sea Basin continued to subside,
and glacio-eustatic sea-level fluctuations, combined with a
high clastic influx, governed sedimentation patterns (4).

From the above, it is concluded that earthquake activities in
the North Sea are probably related to its continued subsidence.
This presumably involves brittle deformation of the upper crust
and ductile deformation of the lower crust and upper mantle.
Whether lithospheric cooling processes, probably involving
phase changes, could induce elastic failure of the upper mantle
and lower crust and consequently give rise to earthquakes is
not known.

6. CONCLUSIONS

As a result of the intensive exploration efforts of the oil
industry, the tectonic framework of the North Sea and of the
adjacent onshore areas is well known. In large parts of the
North Sea area sediments are, however, so thick that routine
seismic surveys, as carried out by the oil industry, are not
able to map the top of the crystalline basement. Moreover,
relatively little is known about the crustal configuration of
the North Sea.

The location of the epicentres of historic earthquakes that
occurred in the North Sea are poorly known and their focal
depth even less. This renders it difficult to relate the
earthquake activity of the North Sea to its basement tectonics.

The North Sea area has undergone a long and complex geological
evolution during which periods of intense tectonic activity
alternated with periods of relative tectonic quiescence. At
present, tectonic activity in the North Sea is at a low level

and is limited to regional subsidence in response to cooling of
the lithosphere and its isostatic adjustment to sediment
loading and eustatic sealevel fluctuation. The post-glacial
rebound of Fennoscandia affects also the eastern margin of the
North Sea. The currently tectonically very active Rhine rift
system does not extend into the North Sea.
Reflection seismic data give no evidence for the Holocene
reactivation of faults in the North Sea that involve the
basement. Neogene and Quaternary strata are, however, locally
cut by faults that are induced by the gravitational instability
of sediments (growth faults; diapirism of salt and salt
solution).

In view of the above, geologists familiar with the tectonics of
the North Sea are hard-put to explain its seismicity.
In order to gain a better understanding of the current tectonic
environment of the North Sea area, and of its seismicity, a
multi-disciplinary approach to this problem is warranted. In
view of the more complete data sets available for the onshore
areas, it is necessary to relate their earthquake activity to
their neotectonics, their crustal configuration and their
structural framework. Moreover, an effort should be made to
quantify the Neogene and Quaternary subsidence pattern of the
North Sea and to get a better understanding of its crustal
configuration. These studies should culminate in the
compilation of a seismotectonic map of the North Sea area, on
the basis of which more concrete recommendations for further
research could be formulated.

SELECTED REFERENCES

1.   Caston V.N.D.: 1979. In: The Quaternary History of North
      Sea. Acta. Univ. Ups. Symp. Univ. Ups. Ann. Quing. Cel: 2
      pp. 23-28.

2.   Heybroek, P.: 1974. Geol. Mijnbouw 53, pp. 43-50.

3.   Illing, L.V. and Hobson, G.D.: 1981. Petroleum Geology of
      the Continental Shelf of North-West Europe. Institute of
      Petroleum.
      Heyden and Son. London pp. 521.

4.   Jelgersma, S. : 1978. In: The Quaternary History of the
      North Sea. Acta. Univ. Ups. Symp. Univ. Ups. Ann. Quing.
      Cel.: 2. pp. 233-248.

5.   Ortlam, D.: 1981. Geol. Rundsch. 10 (1) pp. 344-353.

6. Peltier, W.R.: 1980. In: Dynamic of Plate Interiors.
    Geodynamics Series vol. I. Am. Geoph. Union/Geol. Soc. Am.
    pp. 111-128.

7. Zagwijn, W.H. and Doppert, J.W.Chr.: 1978. Geol. Mijnbouw
    57: 577-588.

8. Ziegler, P.A.: 1982. Geological Atlas of Western and
    Central Europe.
    Shell Internationale Petroleum Mij. B.V. Elsevier pp. 130.

NORTH SEA SEISMOTECTONICS AND THE ROLE OF GEOLOGY IN OFFSHORE
SEISMIC RISK ASSESSMENT

Colin A Davenport

Department of Applied Geology,
University of Strathclyde, Glasgow G1 1XJ, UK.

Some aspects of the role of geology in seismic risk
assessment are discussed. Tertiary faulting is believed to be
essentially of normal dip-slip type within poorly consolidated
sedimentary sequences and may well be aseismic. Sources of main
earthquakes are suggested to lie on deeper reactivated crustal
faults, or perhaps even within the uppermost mantle.

The problems presented by the particular, as yet poorly
understood and even perhaps unique, intraplate seismotectonic
regime of the North Sea area are discussed, including potential
for fault reactivation, fault rupture/magnitude relationships
and paleoseismicity.

Perspectives are offerred to encourage rapid research and
bold synthesis,and conclusions and ideas are presented as a
contribution towards an engineering seismotectonic scheme for
the North Sea area.

A. R. Ritsema and A. Gürpınar (eds.), Seismicity and Seismic Risk in the Offshore North Sea Area, 15–30.
Copyright © 1983 by D. Reidel Publishing Company.

1.0 INTRODUCTION

The utilization of geological, tectonic and earthquake data from both offshore and adjacent land areas to produce realistic seismotectonic schemes is increasingly regarded as a pre-requisite for the safe design of offshore structures, power plants, and storage and disposal sites for strategic and hazardous materials. For many years the North Sea area was regarded as a crustal zone of exceptionally low seismic activity. During the past decade our awareness of the historical and present-day earthquake hazard has increased to a point where the substantial investments represented by North Sea technology require the protection of seismic risk assessment. This research workshop at Utrecht is an acknowledgement of the need for an improved multidisciplinary evaluation of the nature of the earthquake hazard in the area and also the need for proposals for improving our understanding of the risk levels.

The seismic data base for the offshore areas of the North Sea region is acknowledged to be poor, at least by comparison with the land areas, and certainly when compared with more seismically active areas where offshore and onshore advanced technology requires the quantification of seismic risks, e.g. Western U.S.A. and the Mediterranean. The principal geological components of earthquake hazard, i.e. active tectonic features, are poorly understood and todate an active structure has not been identified within the North Sea area. Latterday seismic risk techniques have developed a philosophy which requires geological knowledge to augment and even predict seismic phenomena in areas where earthquake data are poor, i.e. a tectonic scheme is used to provide the extent of causative structures, the limits of seismotectonic provinces[1], maximum earthquake values for faults and provinces etc. Although such techniques for utilizing geological data are now used extensively in safety studies for the siting and design of Nuclear Power Plants and kindred facilities, for example see the guidelines/regulations of the United States Regulatory Commission[1] and the International Atomic Energy Agency[2], they contain many uncertainties and controversial elements. To a certain extent the analyses applied to the data can quantify the uncertainties, however, the costs of conservatism required to compensate for inadequate knowledge can be unacceptable and perhaps can result in the adoption of inadequate geological and tectonic models. Our relative lack of knowledge of offshore geology merely compounds the problems. Therefore it is not surprising that a satisfactory seismotectonic scheme has not been published yet for the North Sea area.

The production of a realistic seismotectonic scheme for the North Sea area, however preliminary, should be regarded

as a priority project to be supported by all concerned with this
Workshop. Only by synthesizing current knowledge and presenting
informed interpretations on a map, will engineers be able to
improve risk evaluations and reduce extreme conservatism.
Moreover, such a map would provide a basis for the planning of
research and focus efforts on the relationships between
terrestrial and sub-seafloor structural features and neotectonic[2]
phenomena. To this end, our understanding of the Quaternary
geology and fault mechanics of the North Sea area must be
improved, and at the present time there is no alternative but
to utilize the techniques developed by scientists and engineers
in more seismically active areas of the world.

2.0 GEOLOGICAL ASPECTS OF EARTHQUAKE RISK ASSESSMENT

The relationship between earthquakes and tectonic stresses
associated with major fault movements lies at the heart of
seismic risk evaluation. Most earthquakes are produced by slip
and rock breakage along existing faults at relatively shallow
depths. A seismotectonic synthesis for the evaluation of the
earthquake hazard to a site requires both high quality seismic
and geological data on regional (over an area of several
hundred kilometres radius) and local (within a radius of five to
ten kilometres) scales. Regional investigations provide the
tectonic framework required to group earthquakes into far field
source areas by defining major causative structures (fault
zones)[3] and seismotectonic provinces[1]. Site area (local) studies
are required to be more comprehensive in order to locate smaller
scale tectonic features which may be capable of producing near-
field earthquakes and surface displacements at the site. On a
local scale, geological investigations are required to locate
associated hazards such as karst phenomena, unstable ground and
slopes, and areas of liquefaction potential. These hazards,
along with tsunami waves, result from both larger finite displacements and vibratory ground motions.

Vibratory ground motion parameters used in structural
engineering design are estimated using "design" and "safety"
earthquakes based upon (a) deterministic maximum event values
and (b) a probabilistic treatment of the total contribution of
a number of source-related maximum earthquake values attributed
to structure or province.

Both approaches require the application of a common set of
geological data and for this reason the seismotectonic method
has been developed. Current guidelines governing the application
of this method to safety and design studies for nuclear power
plants are of limited value. Appendix A to USNRC Guidelines 10
CFR 100([1]) does not define seismotectonic provinces nor does it
include details for the investigation of seismicity and geology

in offshore areas; a problem for the siting of coastal and offshore facilities. IAEA Safety Series Guidelines No. 50-SG-31([2])
provides guidance on information, investigations and methods for
evaluating a design basis for earthquakes, including the seismotectonic method integrated into both deterministic and probabilistic approaches. Here again, guidelines on offshore applications
are extremely limited.

The geological contributions can be divided into three
main groups:

(a) definition of seismotectonic province boundaries and
     causative fault zones (source areas).

(b) definition of maximum magnitudes.

(c) definition of frequency of occurrence of earthquakes.

It should be stressed that although these contribuitions
are particularly important in areas of poor seismic data, such
as the North Sea, geologists have only recently acquired
sufficient research experience and techniques to resolve these
issues.

3.0 DEFINITION OF SOURCE AREAS

3.1 Seismotectonic Province Boundaries

The most general seismotectonic map would show global
tectonic plate boundaries, such as collision, subduction and
spreading-ridge zones. The boundaries are plate contacts, i.e.
zones of greatest recurrent seismicity (interplate earthquakes)
and the provinces are the plates themselves, each with a
characteristic level of seismicity (intraplate earthquakes)
somewhat lowever than that of the plate boundary zones. Large
intraplate earthquakes do occur within locally active zones;
zones which may be many kilometres in width and hundreds if not
thousands of kilometres in length.

In plate boundaries, earthquake zones have distinctive
depth distributions, such as the Benioff zones of subducting
plates, and changes in earthquake activity with depth and
relationships to velocity zones can be used as criteria for
separation into seismotectonic provinces.

In seismic risk studies, plate boundaries can be isolated
as causative features, i.e. as very narrow elongated provinces.
The subdivision of intraplate areas on more local scales is more
complex. Any seismotectonic regionalization of the North Sea
area has to take account of its intraplate location, i.e. well

within the Eurasian Plate and far removed from plate boundaries.
Ideally seismotectonic province boundaries should separate
seismicity patterns into distinct areas agreeing with the present
day stressfield; the stressfields often being inferred from
neotectonic information. The geological evidence available
for the North Sea area gives only the most general information
on the neotectonic regime and the lack of reliable and abundant
earthquake case-histories prohibits meaningful zonation, i.e.
based on differences in seismicity rates and source mechanisms.
Therefore, any seismotectonic provinces proposed for the North
Sea area should be regarded as very preliminary.

3.2 Causative Fault Zones

The identification of seismically active fault zones
requires adequate description of hypocentres (in terms of
location, depth and first motion chracteristics), fault geometry
and history. The activity of shallow neotectonic structures
with surface expression can be assessed using geological
criteria similar to those used to assess the capability[4] of
faults for the siting of Nuclear Power Plants. These criteria
are:-

(a)   evidence of single movement within the past 35,000
      years or multiple movements within the past 500,000
      years (often increased to 700,000 years to apply the
      Brunhes -Matuyama paleomagnetic reversal).

(b)   macroseismic activity associated with a fault.

(c)   structural relationships such that movement on a
      capable fault can cause movement on another.

The macroseismic criterion is often difficult to apply
because sufficiently accurate locations are rarely available.
A promising approach to fault activity involves the accurate
location of microearthquakes (magnitudes less than M=3.0) and
the possible interpretation of microearthquake - fault
associations as active features. Here again, sufficient location
accuracy is difficult to achieve principally as a result of
interpretation sensitivity to crustal velocity and depth control.
Costs of fault activity studies are high since they require
special instrumentation networks. The geological criteria are
often applied. They require investigations involving
excavation of fault plane material and undisturbed strata, etc.,
and, being expensive, such detailed studies are normally
restricted to near-site (local) faults.

The direct application of nuclear power plant safety
criteria to offshore investigations is not required. The

activity of offshore faults is best based upon the definition of
Bonilla[3] who proposed that an active fault is one which has
moved within the Holocene Epoch (about the last 10,000 years).
The Holocene conforms to the time period since the end of the
last major continental glaciation and in Europe is generally
considered to be the past 10,000 years ([4,5]). Bonilla[3] felt
that this criterion was adequate for important non-nuclear
engineering structures and his approach has been supported in
general terms by many others, including Allen[6] and Bolt et
al[7].

The age of last movement of some normal faults in the
Central Mediterranean region has been established for offshore
platform safety studies using this criterion with changes in
sedimentation and the ecology of microfossils. However, it
remains to be seen how successful such an approach would be in
the environs of the North Sea.

3.3 Seismotectonic Regime of the North Sea Area

What we should explore now are the possibilities for
proposing seismotectonic provinces and causative fault zones
in the North Sea, at the same time indicating the research
which needs to be carried out before an acceptable scheme can
be adopted. Ziegler[8] has provided the Workshop with a comprehensive synthesis of the evolution of the North Sea area.
There are many faults within the pre-Cretaceous and early
Cretaceous strata and, on the basis of available seismic
stratigraphy, it appears that few of these faults persist into
the late Cretaceous and early Tertiary strata. Tertiary sedimentary rocks thicken dramatically in the central areas of the
North Sea forming the infills of several "failed rift features"
- the so-called axial grabens. The Tertiary is also characterized by high sedimentation rates and subsidence. Although
faulting appears to be rare in the Tertiary sequences, it is
believed that improved resolution and interpretation of seismic
exploration data will indicate the presence of many dip-slip
faults in these rocks. These structures are expected to be
basement controlled and the product of persistent extension in
the region[9]. Improved interpretation of such structures is
being facilitated by the simulation of extension tectonics
using laboratory experiments. Various deformation styles
controlled by the attitude of basement faults and the presence
of salt domes and deltaic rollover anticlines have been
demonstrated by Horsfield([10,11]). While the Tertiary faults of
the northern areas of the North Sea, e.g. Viking Graben, are
most likely to have experienced dip-slip motions, the deeper
fault systems of southern areas appear to trend south-eastwards
into the Dutch/German Rhine Graben; where faults also exhibit
strike slip motions. The Central Graben appears to have struct-

ural affinities with the Viking Graben, and these two grabens, along with the western spur branching towards northeast Scotland, are characterised by pull-apart listric faults apparently with little or no connection with the deeper fault zones[9]. The process of thermal subsidence of axial areas and the overall compliance of the crust in the region may well explain the general lack of propogation of reactivated faults through the Tertiary sediments.

It would, however, seem reasonable to enquire whether some major basement faults have been zones of persistent movement and could be located via Tertiary and neotectonic (Quaternary) features.

Extensive faulting of soft sediments occurs in the offshore deposits of the Gulf Coast Plain of the USA as a result of relatively rapid movements precipated by rapid differential compaction and subsidence. These processes produce significant displacements with low stressdrops, i.e. virtually aseismic growth faults, with little or no connection with "basement" structure. This analogue suggests that there may be little potential for significant earthquake activity associted with the Tertiary of the North Sea, although the potential for faulting, possibly as yet largely undetected, is great. It is therefore suggested that much of the seismicity of the North Sea area probably results from the movements of reactivated deeper faults, a hypothesis as yet untested by focal depth values. Earthquakes originating in the mantle cannot be ruled out and may well be of some importance.

The concept of reactivated fault zones within the lower part of the crust and even perhaps the mantle is attractive for the following reasons:-

(a) the major earthquakes of 1927 and 1931 exhibit slow attenuation, a feature of deeper events.

(b) the epicentre of the Doggerbank 1931 earthquake as relocated by Nielson[12] lies at the intersection of the Outer Silver Pit Fault and the Swarte Bank Hinge Zone to the north of the Sole Pit Inversion[13].

There are three important features in the Doggerbank area which suggest reactivation potential; (i) presence of mobile salt walls; (ii) connection with the Dowsing Fault Zone trending south eastwards towards the Rhine Graben, and (iii) the bending zone evidencedby the Cretaceous inversion, and with the Dowsing and Swarte Bank Fault Zones acting as hinge lines[13].

(c) the 1927 event offshore of Norway lies close to the

faulted eastern margin of the Viking Graben, yet
high quality seismic exploration profiles demonstrate
that the Paleogene (Lower Tertiary) apparently over-
steps the margin without significant deformation.

(d) neotectonic processes, principally rapid axial sedi-
     mentation and post-glacial rebound of adjacent land
     areas.

The distribution of the larger earthquakes is somewhat
scattered[14] yet, even allowing for poor network coverage,
several weak patterns are suggested:-

(a) the largest events are offshore and during the past
     half of the century perhaps offshore activity has been
     greater than on land.

(b) clusters of activity occur offshore of southern
     Norway and in the vicinity of the Viking and Rhine
     Grabens, also

(c) northwest of the Viking Graben, possibly along the
     northward extension of the Great Glen Fault of
     Scotland.

From the above considerations, it is clear that we have
some way to go before we shall be able to define the active
(reactivated) fault zones of the North Sea area. Some of the
methods which could be used to aid in the identification of such
features are suggested below:-

(a) the identification of bending zones and their patterns
     of migration with time. Evidence from the Mississippi
     Embayment which has axial and hinge bending zones,
     suggest that bending zones may be considerably more
     mobile in time than the locus of maximum sedimentation.
     In the Mississippi Embayment the axial zone is a
     persistent bending zone and is considered by Krinitzsky
     [15] to have great potential for continuing reactiv-
     ation and earthquake generation. Features such as the
     Sole Pit inversion (possibly responsible for the 1931
     Doggerbank earthquake) are associated with bending
     zones.

(b) the placement of arrays of seafloor seismometers such
     as those developed by the Institute of Geological
     Sciences UK[16] and Sandia Laboratories, USA[17].
     Where abundant records of M<3.0 events (microearth-
     quakes) could be used to provide hypocentral/depth
     information and perhaps define fault zones. It could be

suggested that the main faults in the Sole Pit area
of Doggerbank be given priority for evaluation in
this manner.

(c)   detailed mapping of deep fault zones using high
      resolution seismic reflection interpretation
      techniques to define segmentation and relationship
      to higher-level structures.

(d)   as per (c) to define Tertiary tectonic features, and
      if possible, neotectonic features. The results could
      be correlated with the distribution of surface features
      such as pock-marks(which may result from seismically-
      induced degassing of surficial sediments[18], fault-
      scarps, lines of slides etc. Whilst such features
      would probably indicate essentially aseismic surface
      faulting in the shallow sediments, they would assist
      in the location of deeper seismic ruptures.

(e)   the study of onshore potentially-seismic structures.
      Only recently have extensive field investigations of
      neotectonic movements been carried out in the areas
      around the North Sea. Recent Norwegian studies have
      located seven sites of possible neotectonic fault
      movement[19] and earthquake activity on the major fault
      zones in Scotland, such as the Great Glen Fault and the
      Highland Boundary Fault has been documented for many
      years. Research into the activity of these old fault
      zones should attempt to trace them into the offshore
      basement, since they could be responsible for off-
      shore earthquakes if reactivated. Quaternary tectonics
      and microseismicity are aspects most likely to be
      researched in the near future.

(f)   modelling of aseismic (creep) and seismic (rapid slip)
      motions on gravity dip-slip faults in basement -
      soft cover situations of intraplate type (compare with
      recent reviews of plate boundary models, e.g. Mavko
      [20]).

Because the definition of causative features is not yet
possible for the North Sea area, it is recommended that a preliminary seismotectonic map should be prepared using seismotectonic provinces defined by pre-Tertiary structural associations and fault trends. A suggested group of provinces are
given in Table 1.

TABLE 1: Preliminary Seismotectonic Provinces of the
North Sea area

Seismotectonic Province	Pre-Tertiary Fault Trend
1. Coastal Norway	NE - SW, N - S
2. Viking Graben	N-S
3. Shetlands and Buchan	NE - SW
4. Central Graben	NW - SE, N - S
5. West Central North Sea [Scotland, England]	NE - SW
6. East Central North Sea [Denmark]	NNE - SSW
7. Rhine - Anglian	NW - SE

4.0 DEFINITION OF MAXIMUM MAGNITUDES AND FREQUENCY-MAGNITUDE RELATIONSHIPS.

Of the various methods available for estimating the
maximum magnitude for a source, there are two quantitative
approaches which can be based upon geological information:-

(a) fault rupture-magnitude data.

(b) frequency-magnitude relationships, particularly where
historical and instrumental records of earthquakes are
few and conditions favour the use of palaeoseismic
information.

4.1 Fault rupture - magnitude relations

Prediction criteria for earthquake magnitude values versus
surficial fault rupture dimensions have been presented by Tocher
[21], Press[22], Mark and Bonilla[23] and Slemmons[24]
[fault length values]; Bonilla and Buchanan[25] [fault displace-
ment]; and Utsu and Seki[26], Kanamori and Anderson[27], Wyss
[28] and Singh et al.[29] [fault area]. Many of the field
observations were derived from Californian strike-slip faults.
In all cases the rupture dimensions are assumed to be the result
of the earthquake generation process.

When applied to the seismotectonic regime of the North Sea,
the built-in assumptions and difficulties of application become
clear:-

(a) although rapid fault slip occurs during the co-seismic
phase of activity, there may be individual and variable

components of deformations within the pre-seismic
phase. Post-seismic creep may be initiated in the co-
seismic phase and have produced a large percentage of
the overall displacement by the time the succeeding
interseismic strain-accumulation phase commences.
Therefore a large and variable proportion of surface
rupture displacement could be aseismic strain, a
situation most likely in the North Sea sedimentary
sequences. Moreover, a seismic slip may not actually
reach the surface and therefore cannot be characterised by surface measurements.

(b) Ambraseys and Tchalenko([30]) produced a magnitude/
surface rupture envelope for worldwide faulting,which
demands a lower limit (cutoff) of M = 5.4, below which
surface rupture has no significance with respect to
magnitude. Many of the largest North Sea earthquakes
have estimated magnitude values which do not exceed
this cutoff value. Ringdal([14]) suggests that the
largest recorded events may have been M = 6.0 to 6.5
and that maximum earthquake levels could be M = 7.0 to
7.5. If we assume a design event of M = 7.0 for a
return period of several hundred years, published
surface rupture data indicate possible fault lengths
exceeding 100 kilometres and displacements of the
order of 1.5 metres or more. If sufficiently fresh,
such displacements and the associated sediment
deformations due to water and gas expulsion, could be
detected on sonargraphs and echotraces particularly
in marginal areas where better consolidated materials
occur at or close to the surface.

The detection and quantification of the morphological
evidence required to define displacements would be difficult
under North Sea conditions except perhaps on rocky platforms
off the coasts of Norway and Scotland. Resultant fault scarps
could be evaluated using paleoseismic[5] techniques, such as
those described by Wallace([31],[32]). Wallace finds that, in the
terrestrial environment, the form and size of fault scarps in
unconsolidated materials can reveal the ages and moment magnitude of prehistoric earthquakes. This is particularly true of
tectonic domains chracterised by normal faults. Much of the
work carried out for this approach requires the drilling of
boreholes for samples, trenching, age-dating and careful
logging and mapping.

In general, time prediction and slip prediction models do
not apply since similar slip increments estimated from scarps
and colluvial wedges are often separated by very different time
intervals. Therefore a characteristic earthquake model is

indicated.

4.2 Frequency - Magnitude Relationships

Frequency-magnitude relationships can be used to predict maximum earthquakes, and are also required for each seismotectonic province in a total-contribution probablistic analyses for earthquake risk. The frequency with which prehistoric earthquakes have occurred can be estimated from the mapping of deformed sediment layers[33]. Sims has described small-scale liquefaction structures in reservoir sediments exposed after the failure of the Van Norman dam during the M = 6.6 San Fernando Valley, California, earthquake of 1971. Since then comparable structures have been described from Canada[34] in contorted layers of late-glacial varve deposits which have been correlated over distances of 160 kilometres, and also from Turkey[35] where lacustrine sediments close to the East Anatolian Fault yield similar features. Sims has estimated that such sediment deformation would only take place when the local intensity of ground motion reached VI on the Modified Mercalli Scale. Therefore, the frequency of strong events capable of liquefying and deforming sediment layers, can be estimated by age-dating lake sediments or by counting varve couplets. Estimates based on these techniques could provide minimum recurrence values exceeding those based upon data from other sources, i.e. instrumental and historical. From the seismotectonic considerations presented herein, it would appear that there is considerable potential for the application of palaeoseismic techniques in onshore evluations of the North Sea area.

Paleoseismic investigations are required in all countries around the North Sea. Particularly promising areas for research are the northern parts of the United Kingdom and Scandinavia.

5.0 CONCLUSIONS

The stated objective of this paper was to discuss the complex role of geology in offshore seismic risk assessment and to provide positive guidance on the key issues which must be placed into perspective if a satisfactory data base is to become available in the near future. The issues focus upon the preparation of a suitable seismotectonic scheme and the necessary source parameters, maximum earthquake values and frequency-magnitude relationships.

Of immediate concern is our apparent inability to model intraplate tectonics and, particular, the mechanisms of motion and earthquake capability of normal dip-slip faults in a predominantly extension regime.

Our understanding of the tectonics and geological history of the area suggests that at least the larger earthquakes are produced by movements on reactivated deep faults within the crust and possible even within the mantle. Improved coverage of modern seismological instrumentation, both on land and on the seafloor, is required to provide satisfactory focal locations and first motion parameters, thereby testing the reactivation hypothesis.

Until seafloor microearthquake networks are deployed, any fault activity will probably remain undetected. Neotectonic features have not been positively identified on the seafloor and although the indications are that surface rupture and associate phenomena may be rare and of limited scale, more rigorous investigations are recommended. In the meantime, a seismotectonic province approach to risk evaluation is recommended, and a preliminary list of provinces has been presented.

Meaningful paleoseismic data are most likely to be obtained on land and increased research efforts are recommended, particularly to identify any fault activity and prehistoric earthquake records in late glacial sediments associated with fault zones which trend offshore with potential for reactivation.

NOTES:

1. SEISMOTECTONIC PROVINCE - a geographical area characterised by similarity of geological structure and earthquake characteristics.

2. NEOTECTONIC - for seismic regions, the tectonics of the Quaternary period[2]. (circa 1.85×10^6 years before present[36]).

3. CAUSATIVE STRUCTURE - tectonic structure of fault zone known to be producing earthquakes.

4. CAPABLE FAULT - a fault which has significant potential for relative displacement at or near the ground surface[2]. Because of the nature of strike-slip faulting in California, also capability for earthquake generation[1]. The term "active" can be used to describe aseismic as well as seismic faulting.

5. PALEOSEISMICITY - the identification and study of prehistoric earthquakes by the use of analyses of microstratigraphic relations along faults, fault scarp and terrace morphology, and seismically-induced sedimentary structures.

REFERENCES

(¹) United States Nuclear Regulatory Commission (USNRC), 1973,
     "Nuclear Power Plants - Seismic and Geological Siting -
     10 CFR 100, Appendix A", Federal Register of the United
     States of America.

(²) International Atomic Energy Agency (IAEA), 1979, "Earth-
     quakes and Associated Topics in Relation to Nuclear
     Power Plant Siting - A Safety Guide", Safety Series No.
     50 - SG - S1, Vienna.

(³) Bonilla, M.G. 1970, "Surface faulting and related effects",
     in Earthquake Engineering, ed. R.L. Wiegel, Prentice-
     Hall, Inc., U.S.A. chpt. 3, pp. 47-74.

(⁴) Flint, R.F. 1974, "Glacial and Quaternary Geology",
     John Wiley, New York.

(⁵) Van Eysinga, F.W.B., 1975, "Geological Timetable",
     Elsevier, Amsterdam.

(⁶) Allen, C.R., 1975, "Geological Criteria for Evaluating
     Seismicity", Bull. Geol. Soc. Am., 86, pp 1041-1057.

(⁷) Bolt, B.A., Horn, W.L., MacDonald, G.A. and Scott, R.F.,
     1975, "Geological Hazards," Springer-Verlag, Berlin.

(⁸) Ziegler, P.A., 1981, "Evolution of Sedimentary Basins in
     North-West Europe", Petroleum Geology of the Continental
     Shelf of North-West Europe, Institute of Petroleum,
     Chpt. 1, pp. 3-39.

(⁹) Gibbs, A.D. 1982, "Panel contribution: Session 1 - Geology
     and Tectonics". NATO Advanced Research Workshop on the
     Seismicity and Seismic Risk in the Offshore North Sea
     area, Utrecht, (this volume).

(¹⁰) Horsfield, W.T., 1977, "An experimental approach to
     basement-controlled faulting", Geol. en. Mijnbouw, 56(4),
     pp. 363-370.

(¹¹) Horsfield, W.T., 1980, "Contemporaneous mvoement along
     crossing conjugate normal faults", Journ. of Structural
     Geol., 2 (3), pp 305-310.

(¹²) Nielsen, G. 1979, "Historical Seismicity of the North Sea",
     British Assoc. Advancement of Science Annual Meeting,
     Section C., No. 74.

([13]) Glennie, K.W. and Boegner, P.L.E., 1981, "Sole Pit Inversion Tectonics", *Petroleum Geology of the Continental Shelf of North-West Europe, Institute of Petroleum, London,* Chpt. 9, pp. 110-120.

([14]) Ringdal, F., 1982, "Seismicity of the North Sea area", NATO Advanced Research Workshop on the Seismicity and Seismic Risk in the Offshore North Sea areas, Utrecht (this volume).

([15]) Krinitzsky, E. J. 1974, "State of the Art for assessing earthquake hazards in the United States". Report 2: Fault Assessment in Earthquake Engineering", Miscell. Paper S-73-1, Report 2, National Technical Information Service, USA.

([16]) Anon, 1979, "UK and Norway Study North Sea Earthquakes", *Offshore Engineering,* November, pp. 71-72.

([17]) Anon, 1982, "Earthquake Measuring device aims for better platform design", *Oil and Gas Journal,* April 26, p.86.

([18]) Fannin, N.G.T., 1980, "The use of regional geological surveys in the North Sea and adjacent areas in the recognition of offshore hazards", *Offshore Site Investigations,* ed. D. A. Ardus, Graham & Trotman, London, pp. 5-35.

([19]) Norwegian Geotechnical Institute (NGI), 1981, "Neotectonic movements in Norway", Internal Report No. 40009-7.

([20]) Mavko, G.M.,1981,"Mechanics of Motion on Major Faults", *Ann. Rev. Earth Planet Sci., 9,* pp. 81-111.

([21]) Tocher, D. 1958, "Earthquake Energy and ground breakage", *Bull. Seism. Soc. Am., 48,* pp. 147-153.

([22]) Press, F., 1967. "Dimensions of the Source Region for Small Shallow Earthquakes", VESIAC Report 7885-1-X pp 155-164.

([23]) Mark, R.K. and Bonilla, M.G. 1977, "Regression analysis of earthquake magnitude and surface fault length using the 1970 data of Bonilla and Buchanan", *U.S. Geol. Surv. Open-file Report No. 77-614.* 10p.

([24]) Slemmons, D.B. 1977, "State of the Art for assessing earthquake hazards in the United States : Report 6, Faults and Earthquake Magnitude", Miscell. Paper S-73-1. Report 6, prepared for the Office of the Chief of Engineers, U.S. Army, Washington, D.C.

[25] Bonilla, M.G. and Buchanan, J.M. 1970, "Interim Report on Worldwide Historic Surface Faulting", *U.S. Geol. Surv. Open-File Report,* 32p.

[26] Utsu, T., and Seki, A., 1954. "A relationship between the area of aftershock region and the energy of main shock", *Journ. Seism. Soc. Japan, 7,* pp. 233-240.

[27] Kanamori, H and Anderson, D.L., 1975, "Theoretical basis of some empirical relationships in seismology," *Bull. Seism. Soc. Am., 65,* pp. 1073-1095.

[28] Wyss, M., 1979, "Estimating maximum expectable magnitude of earthquakes from fault dimensions", *Geology, 7,* pp. 336-340.

[29] Singh, S.K., Bazan, E. and Esteva, L., 1980, "Expected Earthquake Magnitude from a Fault", *Bull. Seism. Soc. Am. 70,* pp 903-914.

[30] Ambraseys, N.N. and Tchalenko, J., 1968, "Documentation of Faulting Associated with Earthquakes", Report, Department of Civil Engineering, Imperial College, London.

[31] Wallace, R.E., 1977, "Profiles and ages of young fault scarps, northcentral Nevada", *Bull. Geol. Soc. Am., 88,* p. 1267 - 1281.

[32] Wallace, R.E., 1982, "Fault Scarp Analysis in Paleo-seismicity", *Trans. Am. Geophys. Union, 63 (18),* T31-1, p. 435.

[33] Sims, J.D., 1982, "Earthquake-induced deformation in sediment as a paleoseismic indicator", *Trans. Am. Geophys. Union, 63 (18),* T31-6, p.435.

[34] Adams, J., 1982, "Deformed Lake Sediments record pre-historic earthquakes during the deglaciation of the Canadian Shield," *Trans. Am. Geophys. Union, 63(18)* T31-7, p. 436.

[35] Hempton, M., 1982, "Earthquake-induced sedimentary deformational structures, East Anatolian Fault, S.E. Turkey", *Trans. Am. Geophys. Union, 63 (18),* T31-8,p.436.

[36] Ryan, W.B.F., 1973, "Paleomagnetic Stratigraphy" in *Initial Reports of Deep Sea Drilling Project,* ed. A. G. Kaneps, XIII (2), chpt. 47.2.

CONSTRAINTS ON SEISMO-TECTONIC MODELS FOR THE NORTH SEA

Alan D. Gibbs

British National Oil Corporation

The detailed evolution of the North Sea and the probable genesis of the major tectonic units has been graphically described by Dr. Ziegler (this volume).   Several points arise from this overview which are relevant to present day seismicity which need to be emphasised.

Firstly the longevity of many of the major tectonic elements is important in understanding the evolution of the area.   Many of the structural trends evident in the Mesozoic and Caenozoic history of the basin have a Palaeozoic or even possible Precambrian origins.   Confusion in understanding these elements can arise as their early displacement history may be very different from the palimpsest displacements superimposed at a later stage.   Dr. Ziegler, for example, emphasises the strike-slip nature of some of the Hercynian elements generated in the southern North Sea and their subsequent reactivation as strike-slip systems during the Tertiary inversion.   The function of these elements as dip-slip fracture systems during the Mesozoic is arguably more important in the evolution of the basin, particularly further to the north.

The recognition of the Dutch Central Graben and Rhine Graben systems as analogues of the Viking Graben can be misleading as the former have only small finite extension across the basin and their evolution is dominated by steeply dipping strike-slip faults and associated pull-apart rifts.   The Viking Graben, Central Graben and to a lesser extent Buchan Grabens of the North Sea in contrast are dominated by very large extensions on major listric faults which flatten towards a mid-crustal detachment. The resulting minor structures and distribution of seismicity

31

A. R. Ritsema and A. Gürpınar (eds.), Seismicity and Seismic Risk in the Offshore North Sea Area, 31—34.
Copyright © 1983 by D. Reidel Publishing Company.

on such faults is, therefore, significantly different from that observed in the wrench dominated graben to the south . Unfortunately this difference is not, at present, demonstrated by seismicity records as the North Sea lacks focal depth control on epicentres, but must be incorporated in any seismo-tectonic model.

In addition to this change in basement structural style to the north, multiple detachments have been identified in places. These are normally associated with halokinesis in the southern North Sea (Gibbs, 1982). In certain areas there is evidence that such secondary detachments within the Zechstein can occur outwith and on the margins of areas with recognised salt domes or walls. Any seismo-tectonic model would necessitate the inclusion of these secondary detachments as possible seismic source, as well as the mid and possible deep crustal sources.

Dr. Ziegler; op cit) has pointed out the effect of thermal subsidence on the North Sea Basin. Again the degree of subsidence increases in a regional sense to the north and is essentially lacking in the onshore graben of northern Europe. The extent to which the total observed subsidence can be attributed to a simple McKenzie (1978) type basin model, or whether some element of sub-crustal erosion (Ziegler; op cit) is still a matter of debate. The final resolution of this lies between a tectonic model incorporating a different cross-sectional fault model to that presently proposed and sub-crustal erosion. The implications for modelling a fracture system through which present crustal stresses are dissipated is different in each case, as well as implying different mechanisms for generating the crustal stress field.

A more obvious effect of the thermal subsidence of the basin is that although the basement is extensively faulted and the basement faults define the details of the tectonic elements few of these faults propagate through the thick Tertiary sequences. Indeed, many of the important basement elements are effectively 'dead' at the base Cretaceous seismic marker although there is some evidence that these deeper faults are still seismically active. Exceptions to this do, of course, occur both in the shallow detachment faults associated with halokinesis and with the deep basement faults. The whole fracture system, however, is probably necessary for dissipating present crustal displacements. Scant though the evidence is, existing seismic catalogues of the North Sea area show a correlation with the known basement pattern, and very little with known faults in the thick Tertiary.

The examination of commercial seismic shows very little evidence of neotectonic faulting despite the history of

seismicity, not dissimilar to that of the surrounding land areas.
A number of contributors to this conference have pointed out the
presence of fault-breaks even in Pleistocene sediments on land.
Such fault breaks do not appear to be present over much of the
North Sea area.   The conclusion that continuing displacements
evidenced by seismic records of the North Sea area are
structurally attenuated by compaction, bedding plane slip and
'ductile' deformation of the thick Tertiary sequences seems
inescapable .

Dr. Wood and Professor Ambraseys (this volume) have discussed seismo-tectonic models for the North Sea, both from a
geological and risk analysis view point.   The models presented
and examples discussed are all of regions where the seismicity
is invariably correlated with active fault breaks.   In addition
emphasis has been placed on the stability of structures in
response to finite ground displacement.   Such models seem
inappropriate to the North Sea situation where deeper active
faults do not seem to have a surface expression and the later
displacements are structurally attenuated in the overlying
sediments.   This is a major problem in assessing both hazard
and risk.

The lack of any reliable seismic data and seismo-tectonic
model from areas which are geologically analogous to the North
Sea aulacogene, with its thick, essentially unfaulted sedimentary fill in response to thermal subsidence, suggests that
the approach to risk analysis used in other regions is not
adequate for the North Sea.   In addition the extrapolation of
onshore data into the North Sea basin is also of doubtful
validity, in particular, in the northern areas.   Data from the
North Sea itself is also inadequate in lacking both depth control and magnitude and intensity data for the smaller events.
New instrumentation and sea bed recorders should be used to
improve this situation .

In conclusion an understanding of the seismicity of the
North Sea requires new instrumentation to acquire depth controlled data as well as additional records of small events.
This is essential if the patterns of seismicity are to be related
to a tectonic model which is complex both in plan and cross-
section .   A more complete picture of the tectonic elements than
is presently available is also necessary if a seismo-tectonic
model is to be evolved which will account for earthquake distribution and stress distribution .   Stress fields and neotectonic displacements of the land areas provide important
boundary control for such a model .

If engineering risk analysis (Davenport, this volume) is to
be successful, and not based on the assumption of the North Sea

as seismically 'white', an adequate seismo-tectonic model is necessary.   Before such a model exists the application of inappropriate seismic models which have not been derived from 'McKenzie type' basins are misleading.

Sufficient data for the compilation of a geoseismic map for both the deep and shallower tectonic elements exists in the oil industry, usually compiled in the form of tectonic element maps at, for example, a scale of 1 to 1 million.   Should these maps be made available, they would serve as an adequate tectonic base upon which to compile epicentre and magnitude data from earthquake catalogues in sufficient detail to assist in predicting engineering risk to offshore installations.   This approach has been used successfully elsewhere in the world (Davenport, this volume).   Inclusion of shallow tectonic features along with the basement elements displayed on such maps would aid in delineating those areas which may be at risk from shallow focus events.   A seismo-tectonic model of this nature might prove to be of great value to the industry in the design of offshore installations.

I would like to thank The British National Oil Corporation for permission to publish this contribution and to acknowledge the help of the many people within the Corporation who have contributed to my understanding of the problem.

Gibbs, A D (in press); Balanced cross-section construction from
        seismic sections in areas of extensional tectonics,
        Journal of Structural Geology.

McKenzie; D P, 1978; Some remarks on the development of sedi-
        mentary basins; Earth Plant. Sci Lett, 40, 25-32.

WHAT HAPPENS TO THE RHINE GRABEN 'SUB-PLATE BOUNDARY' WHERE IT MEETS THE S. NORTH SEA?

Robert Muir Wood

Principia Mechanica Ltd/Cambridge

The fate of the Rhine graben zone of seismicity as it
extends to the North West remains a considerable
unknown. From a re-evaluation of the record of British
historical seismicity and evidence for recent faulting
a provisional seismotectonic model is emerging that
can hope to extend the understanding already achieved
for the Rhine graben, through the south North Sea,
and into England. Rapid and differential subsidence
along the S.E. Essex coast supports a model of regional,
tectonically induced, tension.

    The seismicity that extends through the region
to the north and north-west of the Alpine foreland
is almost all restricted to a zone associated with the
Rhine graben system. The amount of seismicity along
the Rhine and its localisation have prompted some
German investigators to term the zone a 'sub-plate
boundary'(for example Baumann, 1981). The geodetically
observed slip-rate for both the Upper and Lower Rhine
grabens is (at approximately 1mm per year) about two
orders of magnitude less than that found at a plate
boundary (Ahorner, 1975). In the Lower Rhine there
is an apparent divide in the zone of seismicity
with one band of activity passing NW into Holland and
the other passing W into Belgium.
    The northernmost of these two bands appears within
the historical record to die out in Central Holland,
but the evidence of a number of recent fault zones,
some showing continued movement, suggests that this

A. R. Ritsema and A. Gürpınar (eds.), Seismicity and Seismic Risk in the Offshore North Sea Area, 35–42.
Copyright © 1983 by D. Reidel Publishing Company.

band continues out into the S. North Sea. The recent
faults reveal continued rifting and estimates have
been made of between 100 - 200m of extension since
the Early Quaternary.

The second band of seismicity, that passes to the
west through Belgium has no accompanying evidence
of surface dislocation and is defined from a relatively
small number of important historical earthquakes.
The band cuts through the Brabant Massif that has
maintained a consistent geological history of
stability. Focal mechanism determinations for events
from this band indicate right-lateral strike slip
motion (Ahorner, 1975). The most recent major event
along this band was that of June 11th 1938, the
isoseismals of which show considerable elongation
along the direction of the presumed fault plane.

Both these bands of activity are shown on all
available maps of seismic zoning and seismotectonics
to run out before reaching the sea. However tectonics
respects neither coastlines nor national frontiers.

When considering intraplate tectonic processes
the geometric constraints on the pattern of movement
are less severe than those implicated in the spherical
geometry of major plate boundaries. This is because the
relatively small amounts of horizontal movement can
become absorbed in vertical crustal deformations
and displacements. If some pattern of intraplate
tectonics is entirely associated with vertical
movements then the accompanying seismicity can be
entirely isolated. Alternatively zones of horizontal
movement may interconnect with zones of vertical
deformation or with plate boundaries.

Since the beginning of the Cenozoic NW Europe
has lain imbetween two plate boundaries, both of
which have been far from ideal in their behaviour.
To the south the Alpine orogeny has involved movements
among a variety of micro-plates caught in the jaws of
a major continent-continent collision zone. Compressional activity has shifted over a wide region and
new ocean crust has been formed as crustal blocks
have rotated within the collision zone. To the
north the North Atlantic Ocean was the final stage
in the development of the Atlantic; the axis of
spreading has shifted a great distance on at least
one occasion and in the midst of the ocean there
is a continuing major asthenospheric upwelling
that provides copious quantities cf basaltic magma
to sustain the island of Iceland above sea-level.
Lying in the midst of these two complex plate boundaries

'sub-plate boundaries' have passed through NW Europe through much of the Cenozoic as a response to differential movement transferred to the Alpine foreland, disturbances of the Upper Mantle associated with the partial subduction in the Alps and the fringes of the Iceland 'plume' etc. At various stages, particularly during the mid-Tertiary, this sub-plate boundary threatened to turn into a plate boundary or at least the zone over which the Mediterranean plate 'boundary' could be said to have been in operation extended into NW Europe. A series of right-lateral NW-SE strike-slip faults of Oligocene age have been traced from Devon through the Irish Sea up to Northern Ireland with displacements of over 20kms claimed for the Sticklepath-Lustleigh Fault in Devon (Jenner, 1981). The normal faults that controlled the sedimentary basins of Southern England and the English Channel became reactivated during the Miocene and in the Isle of Wight now show more than 1km of displacement as thrust faults. The continuing deformation of both the highlands and lowlands of Britain through the late Cenozoic has been proposed but is difficult to corroborate because of the general absence of geological formations by which such activity could be dated.

As the first part of a major research initiative into the seismicity and the seismotectonics of Britain and its surrounds a comprehensive re-evaluation of the historical record of British earthquakes is now nearing completion. The relatively advanced state of the understanding of the seismotectonics of the Rhine graben zone has meant that one of the most important concerns of this research is to 'plug' the tectonics of Britain into Europe. The historical record for the SE corner of England extends back at least 600 years. Within this period there are only two events of a comparable size to the 1938 Belgian earthquake. The first of these is the earthquake of May 21st 1382. The felt area of this event (see Melville, 1982) is similarly elongated to that of the 1938 event and extends from Limburg to Evesham. Damage was caused at London, Westminster, Gent and Liege and at Canterbury the campanile adjacent to the cathedral was demolished. The epicentre appears to have been between Bruges and Canterbury but closer to the Kent coast. This event confirms that the Belgian band of activity continues offshore. The second event is that of April 6th 1580 that was felt much further to the north across England and to the south through the Paris basin than the 1382 event (see Melville, 1982b). Minor damage

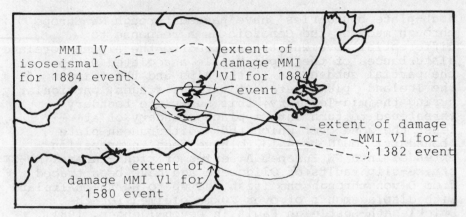

Fig 1. Intensity distributions for crucial SE England
earthquakes.

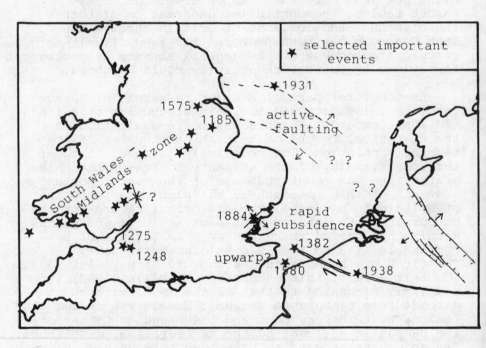

Fig 2.   Postulated seismotectonic sketch map of SE
and E England.

(including two deaths) was caused in London, and more serious structural damage to one or two castles and churches was occasioned around the SE coast of Kent and at Calais where a watch-tower was split in two and houses on the city wall collapsed. Various accounts of sunken ships and 'tidal' waves suggest that this event had an epicentre in the Straits of Dover somewhat to the south of the 1382 event. The damage area of this event shows less elongation than that of 1382 or 1938 forming a NW-SE oriented ellipse with a major axis running from London to Mons and a minor axis from Bruges to Boulogne, see Fig 1.

These two events suggest that the activity and event size of the Belgian zone of seismicity does not decline to the west. However through Southern England there is no continuation of this level of seismicity. The events which have occurred in the counties of Kent, Surrey, Sussex and Hampshire have all been of very small felt area, but sometimes high epicentral intensity, and all associated with the major basement faults known to have controlled the monoclines such as those of the North and South Downs. To find a similar level of activity one must look to South Wales and the extension of the South Wales zone of seismicity into Somerset where the cathedral was damaged in an earthquake of 1248 and the Chapel of St Michael on top of Glastonbury Tor destroyed in an earthquake of 1275 (Flores Historiarum, Rolls Series 95/iii, p46 & Matthew Paris, Chronica majora, v, 46). There are a number of possibilities as to the fate of this zone of strike-slip activity:

1) It continues under Southern England but there has been no major event along it within the past 1000 years.

2) The unexplained inter-relationship between the Belgian and Dutch bands of the Lower Rhine graben zone of seismicity suggest a more complex alternation of phases of activity that could imply that the Belgian zone is relatively young, and still propagating westward.

3) The activity diffuses over a wide region of SE England.

4) The activity somehow becomes diverted to the north and /or the south of Kent and East Anglia.

Should the right-lateral strike slip motion passing through Belgium come to an end close to the English coast then the region to the north-west of this termination should be in extension and the region to the south-west in compression. Some scattered evidence already exists for local tensional faulting in

Central England. Trough faults, involving the deformation
of unconsolidated sediments overlying presumed normal
faults have been described by Shotton (1965). More
specific evidence of crustal extension has come from
the SE corner of the Essex coast where Greensmith and
Tucker (1980) identified 4.3m of differential sinking
since Neolithic (5000 BP) times between Foulness Island
and the nearby Blackwater estuary which they attribute
to movement on a NE-SW oriented monoclinal flexure
that runs through Sales Point. A few kilometres to
the north of this flexure lies the epicentre of the
1884 April 22nd 'Great Colchester Earthquake' that
caused severe damage to hundreds of houses and some
churches (Maximum intensity MMI Vll, locally Vlll-)
in a series of villages to the SE of Colchester.
The energy of this event was thrown to the NW such
that the intensity lV isoseismal extends 125km into
Northamptonshire to the west yet only 35km to the east.
This pattern of directed energy, allied with the local
evidence of geomorphology and known buried fault structures suggests that this event originated on a NE-SW
fault. Where the projected line of this fault crosses
the Stour estuary Shotton (1965) describes from Wrabness
on the south shore of the Stour a trough structure
in the London Clay filled with Stutton Brickearth that
contains mammalian fossils of the Eemian interglacial,
and that is bordered by a vertical slickensided junction.
Evidence for a longer term flexure between East Anglia
and the adjacent North Sea is provided by the difference
in elevation between the base of the Pliocene of some
400m from onshore to offshore deposits, all of which
has developed since the shallow marine deposits were
laid down.

     To the south of the Thames estuary there is some
suggestion that the coast of Kent is not engaged in
the same rapid sinking as the coast of Essex. In
particular the removal of the harbours of Winchelsea
and Rye from access to the sea, and the continuing
extension of the Dungeness shingle ridges would seem
implausible if rates of sinking were similar to those
found at Foulness. However because a record of
past sea-levels can only be preserved through subsidence,
there may be a problem in demonstrating this supposition.

     Thus there exists within the evidence of the limited
available data from historical seismicity, coastal
deformation and recent faulting support for a tectonic
model in which the Belgian zone of strike-slip faulting
extends offshore (the 1382 event) from whence in the
south it becomes involved in thrusting (the 1580 event)

and to the north in extension. The rapid subsidence
of the SE Essex coast is of a similar magnitude to
rates of fault movement detected in Holland and
suggests a comparable tectonic origin, see Fig 2.

The Dutch band of the Lower Rhine graben that as
it tends towards the coast is marked by evidence for
recent faulting but low seismicity has not been
traced offshore, nor are there any events of the size
of the 1382 and 1580 earthquakes that can be considered
to continue the band between the East Anglian and
Dutch coasts. There are however within the historical
record a number of events of indeterminate epicentre
that may have originated off the coast of Norfolk.
Evidence that this Dutch zone of activity does
continue offshore is found in the Sole Pit region
to the north of Norfolk where topographic features
suggestive of buried normal fault displacements
are found in some shallow seismic profiles. It
was within this region that the North Sea earthquake
of June 7th 1931 originated and also the 1958 February
9th event which was felt widely over Eastern England.
Historical and instrumental locations of epicentres
suggest that this Sole Pit region is entirely
disconnected from the N.North Sea seismicity of
the Viking Graben and West and South Norway coastal
regions. If the 1931 event represents an extension
of the Lower Rhine graben zone a problem remains
as to the nature of the activity (aseismic?) through
Central Holland and out to the north. That the
Sole Pit seismicity extends onto land is suggested
by the pattern of post-Cretaceous fracture zones
found along the east coast of Yorkshire and the
existence within the historical record of earthquakes
more damaging than any other in the Midlands and
Northern England: - that described from Lincoln in
1185 'On the mondaie in the weeke before Easter, chanced
a sore earthquake through all the parts of this land,
such a one as the like had not beene heard of in
England sithens the beginning of the world. For stones
that laie couched fast in the earth, were remooued
out of their places, stone houses were overthrowne,
and the great church of Lincolne was rent from the
top downwards' (Holinshed R. Chronicles 2 pp188-9) and
the earthquake of 1575 February 26th that at Hatfield
close to Doncaster,   knocked down a number of houses
and that was felt strongly as far as North Wales and the
Severn Valley (Abraham de la Pryme, History and
antiquities of Hatfield, fol 36a).

Acknowledgements: - The pre-1800 record of British
historical seismicity has been researched by Dr Charles
Melville of Imperial College, London.

References

Ahorner L. 1975, Tectonophysics,29 pp233-249.
Baumann H. 1981, Tectonophysics,73 pp 105-111.
Greensmith J.T. & Tucker E.V. 1980, Proc. Geol. Assoc.
91 pp169-175.
Jenner J.K. 1981 Petroleum Geology of the Continental
Shelf of N.W. Europe,pp 426-431.
Melville C. 1982 Bolletino Geophysica Theoretica e
Applicata, Trieste, (in litt.)
Melville C. 1982b Disasters Magazine (in litt.)
Shotton F.W. 1965 Q.J.G.S., 121, pp419-434.

GEOLOGY AND TECTONICS. PANEL DISCUSSION.
INTRODUCTORY REMARKS.

H.A. van Adrichem Boogaert

Geological Survey of the Netherlands, Subsurface,
Oil and Gas Department.

At the Subsurface Department of the Geological Survey we have
been looking at the North Sea area more from a petroleum
geological than from a neotectonic and seismicity point of view.
But our overall structural knowledge leads to the following
remarks.
1) Fault controlled Late Kimmerian basins, which underwent
Subhercynian and Laramide inversion in the Southern North Sea
area, and that are so aptly described by Mr. Ziegler in his
introductory lecture, are present in the Netherlands on- and
offshore. We distinguish the southern part of the Central North
Sea Graben, the Broad Fourteens Basin, the Central Netherlands
Basin, the West Netherlands Basin and its easterly continuation
the Netherlands Central Graben (elswere called Ruhr Valley Graben)
(Heybroek, 1974), and the western part of the Lower Saxony Basin.
    The tectonic activity in these basins ends at the beginning
of the Tertiary, when the saucershaped subsidence pattern of the
North Sea Basin starts to develop. Then in only one basin, the
Netherlands Central Graben in the southern onshore area,
tafrogenesis is resumed in the Oligocene and this continues till
present day. Most active is the fault (or fault system) bounding
the northeast of this graben, the Peel Boundary Fault.
Locally in the landscape in the province of Noord-Brabant a small
fault scarp can be observed. Further descriptions of the
Cainozoic movements, displayed in sections and maps are shown
in the Explanation to the Geological Maps 1:50.000, sheets
51 O, 52 W and 62, and in the Explanation to the Geological
Overview Maps 1:600.000. Seismicity along the faults of the
Central Netherlands Graben is well documented by the KNMI
observatory.
    The Peel Boundary Fault gradually disappears to the north-

43

A. R. Ritsema and A. Gürpınar (eds.), Seismicity and Seismic Risk in the Offshore North Sea Area, 43–45.
Copyright © 1983 by D. Reidel Publishing Company.

west. First the movement is taken over by various splays, then
then no more a true graben edge can be distinguished. So there
are no indications that this fault or fault system continues
into North Sea area. However, some seismicity observed close to
Leiden may be linked with a fault in trend with the Central
Graben system.
2) Other areas with late to recent movements are around and
above salt domes with salt flowage either still active or with
subrosional effects.

In the northeastern Netherlands well developed salt domes
are known with an expression of salt movement in the reduced
thickness of the Quaternary cover. There, however, no seismicity
has been recorded.

Offshore a detailed seismic survey of two salt domes in the
blocks L4, 5, 7 and 8 has been carried out by the Geological
Survey. Apart from conventional seismic coverage, some high
resolution lines have been shot, and Sparker and Boomer surveys
were recorded as well. The results of this study are set in a
report to the Ministry of Economic Affairs and it is expected
that they will be released in due course. In one of these salt
domes at least 350 m of post-Miocene movement took place. Young
crestal faults can clearly be distinguished on the high
resolution lines. These faults appear not to extend to the
surface but they may reach into the Quaternary deposits. The
Sparker and Boomer records show some deep, filled, Pleistocene
fluvial channels.

Detailed site selection surveys, as also indicated by
Mr. Davenport for the Mediterranean area, appear to be of
importance to identify areas with hazards due to (sub) recent
faults close to the surface.
3) To conclude we would like to draw attention to the remarkable
high rate of sedimentation during the Quaternary.

Offshore, just north of The Netherlands, the average rate
was 1.8 cm/1000 yrs. In the Quaternary we find a rate of 25 cm/
1000 yrs, a more than 10-fold increase. Even if we take into
account that the Tertiary sequence is incomplete and that the
compaction of the Quaternary is still in an incipient stage,
the later part of the Cainozoic shows a marked increase in rate
of sedimentation. One could expect that in places this could
give rise to gravitational sliding along growth faults in areas
of rapid subsidence. It should be noted, however, that in the
Subsurface Department no exemples of such faults and their
seismicity are known.

References:
Heybroek, P., 1974: Explanation to tectonic maps of the
                    Netherlands.
                    Geologie en Mijnbouw, vol. 53, p. 43-50.

Zagwijn, W.H. & C.J. van Staalduinen (editors), 1975:
Geologische overzichtskaarten van Nederland
(1:600.000) met Toelichting.
Rijks Geologische Dienst, Haarlem, 134 p.
Geologische kaart van Nederland 1:50.000 met Toelichtingen:
blad Venlo West (52 W) (1967); blad Eindhoven Oost
(51 O) (1973); blad Heerlen (62 W oostelijke helft,
62 O westelijke helft) (1980),
Rijks Geologische Dienst, Haarlem.

POSSIBLE NEOTECTONIC FAULT MOVEMENTS IN NORWAY

P.B. Selnes

Norwegian Geotechnical Institute, Oslo, Norway

The table is based on literature studies and hearsay, and shows locations where possible neotectonic movements have been reported in Norway. Site investigations carried out at Yrkje indicate possible postglacial movements of 6-13m vertically across a NNE-SSE trending fault. In addition, investigations of possible fault movements are being carried out based on geophysical surveys of upper sediments along the Skitfjord pipeline trail.

REFERENCE

Løset, F., 1981. Neotectonic Movements in Norway, NGI report 40009-7.

A. R. Ritsema and A. Gürpinar (eds.), Seismicity and Seismic Risk in the Offshore North Sea Area, 47–48.
Copyright © 1983 by D. Reidel Publishing Company.

TABLE

Location	Feature	Source	Comments
Ødegården, Bamble 1.	Offset of pothole	Brøgger 1884	Probably blasted away, not possible to find today.
Hardangervidda 2.	Fractures with apparent offsets	Reusch 1901 NGI 1981	No Postglacial offsets are discorvered but Quaternary movements are possible.
Tjeldsundet, Northern Norway 3.	Irregularities in the shoreline system	Grønlie 1922	Detailed investigations by spesialists are necessary to prove this.
Gudbrandsdalen 4.	Fractures with apparent offsets	Werenskiold 1931	Visit to the locality is desireable.
Sandnes 5.	Possible fault movements in Quaternary deposits	Feyling-Hanssen 1966	Further investigations may include drilling with core sampling and different laboratory analysis of the sediments.
Meløy, Northern Norway 6.	Remarkable earthquake sequence	Bungum, Hokland, Huseby, Ringdal 1979. Gabrielsen, Ramberg 1979	Detailed geological mapping is desireable.
Yrkje, near Haugesund 7.	Different sedimental environment across fault	Anundsen	Work in progress.

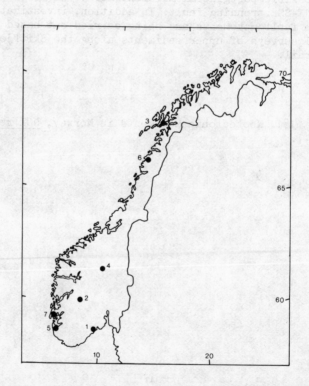

DISCUSSION-NOTES SECTION 1

Geology and Tectonics

Q. P.W. Burton:
   We are told that the North Sea is now tectonically quiescent.
   There is a contemporary strain field in the North Sea; there
   is knowledge of historical earthquakes. It is apparent that
   there are many fault systems of different ages associated
   with different geological epochs in the North Sea. We also
   understand that there is potential for fault reactivation.
   Are there any known North Sea fault systems which can be
   excluded from this potential?

A. P.A. Ziegler:
   To my knowledge there is no evidence from reflection seismic data for the Quaternary reactivation of basement involving faults in the North Sea. However, should movement
   on such faults be very small, they may not be detected by
   routine reflection seismic surveys. Theoretically any basement fault that approximates the orientation of the failure
   planes dictated by a present day stress field could be
   reactivated.

Q. P.W. Burton:
   It seems that if an acceptable tectonic map of the North
   Sea was available then an explanation could be furnished
   for the focal mechanism, focal depth, size, of any historical earthquake on which the seismologist could provide
   data. However, what about tentative solutions to the
   forward problem? Could a tectonic map help to suggest
   which structures might produce earthquakes in the future,
   and perhaps help lead directly towards a seismo-tectonic
   map?

A. P.A. Ziegler:
   In the absence of reflection seismic evidence of faults
   reactivated during the Quaternary it is imperative to know
   as precisely as possible the location of earthquake epicenters and their focal depth in order to relate them to
   possible subsurface structures.

A. C.A. Davenport:
   At the present time focal depth and size parameters for
   any earthquake can only come from instrumental seismic

49

A. R. Ritsema and A. Gürpınar (eds.), Seismicity and Seismic Risk in the Offshore North Sea Area, 49–50.
Copyright © 1983 by D. Reidel Publishing Company.

data. However, it has been necessary to make predictions
from tectonic data for engineering design purposes.
With enough earthquake case history material containing
known depth and location accuracy, we can match earthquakes
with tectonic features to define source areas and causative
faults. Geological knowledge of North Sea tectonics and
fault mechanisms needs to be improved before the focal
mechanisms can be described. In the absence of evidence to
the contrary, it may be reasonable to assume that the
instrumental data-tectonic association will provide a basis
for predicting sources of seismicity.
A suitable tectonic map is the basis of a seismotectonic
map but in itself cannot lead directly to a seismotectonic
scheme. At the present time we are required to anticipate
many possible causative structures in the North Sea area.
My personal view is that the main earthquakes are most
likely to be the result of reactivation movements in deep
crustal fault zones and even perhaps in the uppermost parts
of the mantle.

Q. C.W.A. Browitt:
    In response to a suggestion by Dr. Ziegler that seismologists
    should first address the onshore areas where earthquake
    parameters are better determined, I remarked that there is
    more detailed structural data available offshore. Do you
    agree, therefore, that the problem is one of combining
    earthquake data onshore with detailed geological data
    offshore?

A. P.A. Ziegler:
    The various sedimentary basins of onshore Europe have been
    the subject of continuous exploration efforts for hydro-
    carbons since the Second World War. In addition the density
    of seismological observatories permits a better determination
    of the epicenter location and focal depth of earthquakes in
    the onshore areas as compared to the offshore. Geologically
    speaking the North Sea Basin extends into the onshore areas
    of Denmark, Germany and the Netherlands. Knowledge about
    the epicenter location and focal depth of earthquakes pro-
    vides the interpreting geologist with constraints that
    narrow down his options and that direct his attention to
    specific areas and structural levels. As such the onshore
    areas of the North Sea Basin could prove to be an ideal
    learning ground for the understanding of the seismicity
    of its offshore parts.

Section 2

Seismicity

Invited Paper

SEISMICITY OF THE NORTH SEA AREA *

Frode Ringdal

NTNF/NORSAR, P.O. Box 51, N-2007 Kjeller, Norway

ABSTRACT

The North Sea area is well removed from major tectonic plate
boundaries, and in consequence the seismicity of the region
is modest in comparison to many other areas of the world.
Nevertheless, a few significant earthquakes (M ~ 6.0) are
known to have occurred in the North Sea and adjacent areas
within the past 100 years, the largest of these being the
1904 Oslofjord earthquake, the 1927 earthquake off the west
coast of Norway and the 1931 Doggerbank earthquake. A major
problem in assessing North Sea seismicity is the lack of
adequate instrumental coverage, and only recently can earthquakes in this region be detected at magnitude 4.0 or lower.
Based on the sparse available data, there is still evidence
that the North Sea seismicity level is higher than that of
adjacent land areas, with most of the known earthquakes occurring off the west coast of Norway. There is at present
no evidence to rule out the possibility of future occurrence
of larger earthquakes (M ~ 7.0) in the North Sea area,
although the probability of occurrence of such events must be
rated very low.

---

* NORSAR Contribution No. 320

A. R. Ritsema and A. Gürpınar (eds.), Seismicity and Seismic Risk in the Offshore North Sea Area, 53–75.
Copyright © 1983 by D. Reidel Publishing Company.

INTRODUCTION

The purpose of this paper is to review available data on earthquake occurrence in the North Sea area. In recent years, this
topic has attracted increased attention due to the construction
of numerous offshore platforms and pipelines for oil and gas
extraction. A proper assessment of earthquake hazard is of considerable importance in order to obtain adequate construction
criteria, so as to minimize the chance of failure of such installations.

The structure of this paper is as follows: Following a brief
review of global earthquake occurrence, the remainder of the
paper deals with the North Sea area, and the following topics
are addressed: (i) The data base available for this area,
including macroseismic as well as instrumental data, (ii) Spatial
earthquake distribution in the North Sea and adjacent areas and
(iii) A review in some detail of the largest known historical
earthquakes in the region. The final section gives a brief
discussion on topics such as precision of the solutions, and the
question of the 'largest possible' earthquake in the North Sea
area.

EARTHQUAKE OCCURRENCE - INTER- AND INTRAPLATE EARTHQUAKES

Most of the world's earthquakes occur in narrow bands that can
be clearly identified on a global seismicity map (Fig. 1). With
the emergence of the plate tectonic concept in the 1960's an
attempted unified theory was forwarded to explain the earthquake
occurrence along plate boundaries as resulting from relative
motion of large lithospheric plates. Today, such earthquakes
are commonly called interplate earthquakes, and their mechanisms
and causes are fairly well understood.

A considerably more difficult problem is the occurrence of
earthquakes within plates. Such earthquakes, which are denoted
intraplate earthquakes, occur much less frequently and generally
seem to exhibit less clear tectonic patterns. The North Sea area
is located well within what is called the Eurasian plate, and
is thus an intraplate region. It is clear from Figure 1 that the
earthquake activity in the North Sea is quite modest in a global
perspective, although not insignificant, as will be shown later
in this paper.

Before studying in detail the earthquake occurrence of the
North Sea, it is of interest to summarize some basic facts about
global seismicity, with special emphasis on intraplate earthquakes.
Conceptionally, an earthquake represents a sudden release of accumulated stress along a zone of weakness (or fault) in the crust
or lithosphere (e.g., see Doornbos [1], Knopoff [2], and Madariaga

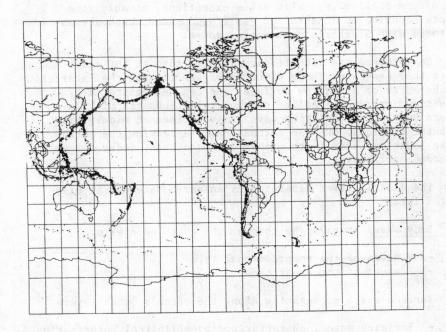

Figure 1.   Seismicity of the world 1967-69 (compiled by
            NOAA, USA)

[3]). Fault sizes may vary considerably, and the largest earthquakes can cause displacement along faults of several hundred
kiolometers' length. While all continents and continental shelves
show an abundance of faults, only very few of these faults are
today seismically active. Thus a geological study is by itself
not sufficient to point out potential areas of seismic hazard.
A further complication is that the large majority of earthquakes,
including numerous destructive ones, do not cause surface ruptures, though imprints of previous movements are clear. In some
cases the entire fault may be buried under thick layers of sediments, making it impossible to be identified through surface
surveys.

     Earthquakes occur at depths ranging from 0 to 700 km.
The deeper earthquakes occur only in lithospheric subduction
zones, and most earthquakes can be classified as shallow, i.e.,
having depths of 70 km or less. Known intraplate earthquakes

are of the shallow type with a few exceptions, notably some
events occuring in Rumania and Pamir/Hindu-Kush which are associated with fossilized subduction zones.

Earthquakes are detected and located through the elastic
waves they generate. These waves propagate through the interior
of the earth or along its surface, and can be registered by
sensitive instruments called seismometers thousands of kilometers away. By associating wave arrival times at widely distributed stations, one can automatically determine the earthquake location, usually to within 15 km accuracy today for well-
recorded earthquakes.

The elastic waves generated by an earthquake can be classified into:

a)   Body waves propagating through the earth's interior

     -   P waves of the compressional type
     -   S waves of the shear type

b)   Surface waves propagating along the earth's outer layers

     -   Rayleigh waves, characterized by elliptical polarization
         in the vertical plane
     -   Love waves, which are polarized in the horizontal plane.

While the P waves are most important for earthquake detection
and location, the shear waves (and in particular the horizontal
component) are of most concern to the engineer.

A number of ways have been proposed to measure the size
of an earthquake. The most widely used measure so far is the
Richter magnitude, which is based on the maximum deflection of
the signals recorded by a seismograph (Richter [4]). The scale
is logarithmic and contains a correction factor which is introduced in order to compensate for the weakening of seismic waves
as they spread away from the source. Recently, an extension of
the Richter scale, applicable to the largest events, has been
introduced by Kanamori [5]. The largest Richter-Kanamori magnitude
yet reported is 9.5, while the most sensitive local seismographs
can record microearthquakes of magnitudes below 0. Most destructive earthquakes have magnitudes above 6, but there are examples
of very shallow, low magnitude shocks which have had disastrous
effects, such as the 1960 Skopje earthquake (M = 5.5) and the
1960 Agadir earthquake (M = 5.8), which each claimed thousands
of human lives.

Earthquakes cause destruction in several ways. Ground vibrations can shake structures and bring them to the point of failure

and collapse. Certain kinds of soil lose their rigidity and
'liquefy' when subjected to seismic ground motions. Avalanches,
mudflows, fire and tsunamis may accompany earthquakes, and sometimes cause much greater damage than the shaking itself.

NORTH SEA SEISMICITY - DATA BASE

Before going into detailed discussion on the earthquake occurrence,
it is important to be aware of both the basic types and limitations
of the observational data available. In general we differentiate
between two types of seismicity data: i) macroseismic data which
reflects how people actually felt an earthquake and ii) instrumental data, i.e., recordings of elastic waves generated by earthquakes using specifically designed instruments, such as seismometers
and accelerometers. Both of these types of seismicity data are
important, and their relative merits with reference to the present
study will be discussed in the following.

Macroseismic data

Macroseismic data simply mean written descriptions on how people
living in the area where the earthquake took place actually felt
the earthquake and, equally important, observable damage to
buildings, bridges, roads, etc.

The systematic collection of macroseismic data on earthquake
occurrence began late in the 19th century, when use of questionnaires was initiated in many countries. Macroseismic data prior
to that time had to be extracted from newspaper reports and
other often very obscure sources and are therefore less reliable,
although the large earthquakes are affected less in this respect.

Several catalogues of historical earthquakes have been
compiled over the years in the countries surrounding the North
Sea, and we mention for Norway Keilhau [6], Kolderup [7], for
Denmark Lehman [8], for the United Kingdom Davison [9], Dollar
[10], for the Netherlands Ritsema [11], and for Belgium van Gils
[12]. An excellent review of available earthquake data for all
of Europe can be found in Karnik [13,14].

Only after around 1800 does the written documentation appear
sufficient to consider reported earthquakes as tectonic events
from a scientific point of view. For example, the earliest reported earthquakes are often coincident with very stormy weather,
which in turn appears to be the real cause of the reported dam-
ages. This tendency of coupling earthquake occurrence with other
geophysical phenomena like storms, thunderstorms, auroras, particular celestial constellations, etc., is well exemplified in
Horrebow's [15] description of the large 1759 earthquake.

The available macroseismic information for an earthquake is primarily used for estimating maximum intensity and radius of perception. The intensity parameter is a measure of the size of the earthquake in question and in this respect the 12-graded modified Mercalli (MM scale is commonly used (Wood and Neumann [16]). Details on the MM-scale are given in Table 1 and demonstrate that the intensity parameter has no obvious physical meaning. Furthermore, several empirical relations have been proposed from which typical earthquake parametes like epicenter location, focal depth and magnitude are related to the basic macroseismic parameters intensity and radii of perception (Båth [17], Karnik 13]).

I.	Not felt
II.	Felt by persons at rest or on upper floor.
III.	Hanging objects swing. Light vibration.
IV.	Vibration like heavy truck. Windows and dishes rattle. Standing cars rock.
V.	Felt outdoors. Sleepers wakened. Small objects fall. Pictures move.
VI.	Felt by everybody. Furniture displaced. *Damage:* broken glassware, merchandise falls off shelves. Cracks in plaster.
VII.	Felt in moving cars. Loss of balance while standing. Church bells ring. *Damage:* broken chimneys and architectural ornaments, fall of plaster, broken furniture, widespread cracks in plaster and masonry, some collapse in adobe.
VIII.	Steering trouble in moving cars. Tree branches broken off. Cracks in saturated soils. *Destruction:* elevated water tanks, monuments, adobe houses. *Severe to mild damage:* brick construction, frame houses (when unsecured to foundation), irrigation works, embankments.
IX.	"Sand craters" in saturated silty sands. Landslides. Cracking of ground. *Destruction:* unreinforced brick masonry. *Severe to mild damage:* inadequate reinforced concrete structures, underground pipes.
X.	Widespread landslides and soil damage. *Destruction:* bridges, tunnels, some reinforced concrete structures. *Severe to mild damage:* most buildings, dams, railway tracks.
XI.	Permanent ground distortion.
XII.	Nearly total destruction.

Table 1.   Modified Mercalli Scale (abridged).

Despite the arbitrariness and subjectivity tied to the collection and analysis of the macroseismic data, this type of information is highly esteemed on two accounts, namely, in providing continuity of observational data over a time interval of several hundred years and in making possible a reasonable estimate of the location and magnitude of the earthquakes in question. The value of historical records of earthquakes is discussed in detail by Ambraseys [18]. However, as far as offshore earthquakes

are concerned, an important shortcoming of macroseismic data is
the difficulty in obtaining reliable location estimates. This
is further discussed in the next section, but it is clear that
macroseismic observations in the majority of cases cause locations
to be strongly biased toward epicenters on land, even for offshore
events.

Instrumental data

Instrumental observations of earthquakes began around 1900 when
the first mechanical pendulum seismographs were installed in
several countries. Due to their low magnifications (around 400
and 200), these instruments did not contribute much detailed
information about the seismic activity in the area under consideration. However, in the period 1955-1965 the seismograph
network was expanded and the instrumental quality vastly improved by installation of modern, high-gain electromagnetic
seismographs with magnification ranging from 15,000 to 150,000.
In 1971 another generation of instruments was introduced with
the large aperture Norwegian Seismic Array (NORSAR) in south-
eastern Norway (Bungum et al [19]). More recently, a number of
local seismograph networks have been installed in the countries
surrounding the North Sea, supplemented by a few ocean-bottom
seismographs, and the coverage of the North Sea area today is
very much improved as a result of these deployments.

     The shortcomings of the local seismograph networks are the
large station separations, the poor azimuthal coverage for earthquakes in the North Sea, and the difficulties in obtaining large
numbers of original seismograph records (which are stored in
different countries). With regard to the latter, it means that
the crucial phase identification or interpretation of the seismic
records is often based on second-hand information, i.e., as given
in local seismic bulletins. Another serious problem is that of
discrimination between the relatively few (tectonic) earthquakes
and the very many artificial events from quarry blasts, naval
activities in the adjacent seas, etc. The above shortcomings of
the seismograph network are, however, not critical for earthquakes
with magnitudes greater than 4.5-5.5, because these events are
also recorded by stations outside Fennoscandia and usually may
be reliably distinguished from man-made explosions.

SEISMICITY MAPS OF THE NORTH SEA AREA

In the previous section, it was stated that the quality of
available earthquake locations in the North Sea must be rated
poor, at least up to the past 10-25 years. Nevertheless, in view
of the relatively modest seismic activity in this region, it is
important to use the earlier available data in order to get
as good a picture as possible of the seismicity of the region.

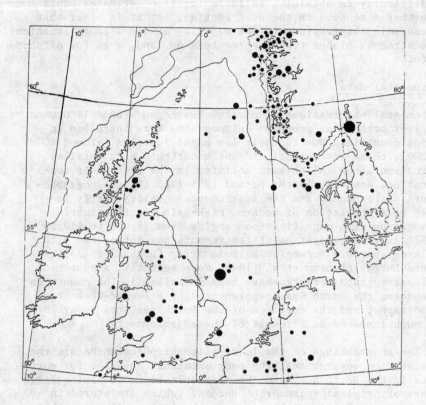

Figure 2.   Known earthquake of M>4 in the North Sea and surround-
            ing land areas as compiled by NTNF/NORSAR and IGS.
            Symbol size corresponds to magnitude, with the
            smallest symbols for M=4.0-4.9, the intermediate
            symbols for M=5.0-5.9 and the largest symbols for
            M>6.0.

We present in the following seismicity maps based on the earthquake catalogues compiled by the Institute of Geological Sciences
- IGS (UK) and NTNF/NORSAR (Norway).

      Figure 2 shows known earthquakes of M=4 and above for the
North Sea and immediately surrounding land areas. We note that
the offshore seismicity appears scattered with a few exceptions,
notably a distinct earthquake zone off the coast of Norway.

Otherwise, it might appear that land seismicity is generally
higher than offshore, but this is clearly a premature conclusion in view of the difference in detectability on- and offshore.
This is well illustrated in Figure 3, showing Fennoscandian
earthquakes from 1951-1980 according to Ringdal et at [20].
While the total number of earthquakes in this map (all magnitudes) is much greater for onshore than for offshore locations,
we see that almost all of the larger earthquakes (M>4) have
in fact occurred offshore. Thus, we may conclude that offshore
seismic activity, at least for the Norwegian continental shelf,
is significantly higher than onshore. The sparse data available
also support the assertion that the regions within the North
Sea with the highest seismicity levels are those immediately
offshore Norway, with the possible addition of an earthquake zone
surrounding the epicenter of the 1931 Doggerbank earthquake.
Simulation studies conducted by Ringdal et al [20] have indicated that the detection threshold for North Sea earthquakes
is typically around M=4.0 for the time period 1951-1980. (At
present, it is probably slightly better.) For the years 1900-
1950, the detection threshold is probably around M=5.0, whereas
during 1800-1900, only events of M>5.5 would likely have been
reported. This poor detectability makes it difficult to obtain
a reliable frequency-magnitude distribution of earthquakes in
the North Sea. Figure 4 shows such a distribution based on the
NTNF/NORSAR catalogue for the years 1951-1980. In viewing this
figure, it should be remembered that the main contribution comes
from earthquakes immediately offshore Norway, and that seismicity
distribution in the North Sea is not homogeneous.

The focal depths of North Sea earthquakes are in general
very difficult to determine accurately, in view of the sparsely
available instrumental data. For the surrounding land areas, use
of macroseismic depth determination techniques can be applied,
and Figure 5 shows the depth distribution of Fennoscandian earthquakes as determined by Ringdal et al [20]. We see that typical
depths range from 5-30 km, and instrumental analysis of selected
well-recorded earthquakes in southern Scandinavia and in Meløy,
N. Norway, have given results within the same range (Bungum et
al [21], Bungum and Fyen [22]). The applicability of these results
to the North Sea area is of course an open question, and the determination of focal depths for North Sea earthquakes remains a highly
important problem for future studies.

THE LARGEST EARTHQUAKES

Of special interest in any seismic risk study are the largest
earthquakes known to have occurred in the region under consideration. Both instrumentally recorded events and earthquakes for which
we have only historical records are of importance here. The 'size'

1951–1980

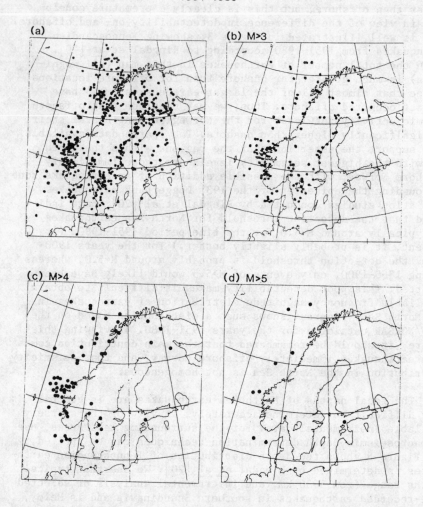

Figure 3. Seismicity map for Fennoscandia and adjacent seas
           covering the interval 1951–1980 and mostly based on
           instrumentally detected earthquakes.

           a) All earthquakes in the NTNF/NORSAR catalogue
           b) Earthquakes of magnitude > 3
           c) Earthquakes of magnitude > 4
           d) Earthquakes of magnitude > 5

           Note in particular that amost all of the larger
           earthquakes have occurred offshore.

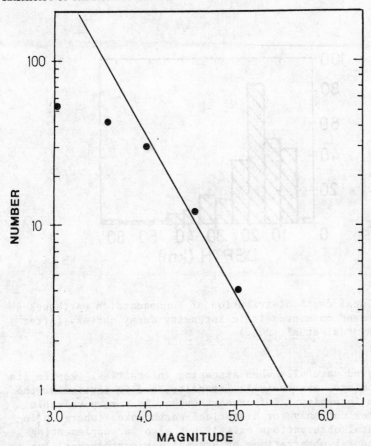

Figure 4. Cumulative frequency-magnitude statistics for the
           North Sea and adjacent coastal areas for the years
           1951-1980. Note the falloff in slope below M=4.0,
           indicating poor detectability at these magnitudes.

of the earthquakes must be determined on the basis of a number of
observed characteristics, and the most prominent in use are:

(a) Macroseismic intensity

     This quantity reflects how human beings have experienced an
     earthquake. For assigning a specific intensity to a given
     observation, the Modified Mercalli scale, which was already
     presented in Table 1, is most commonly used today. As noted
     earlier, intensity is a highly complex quantity reflecting
     mostly ground acceleration but also ground displacement and
     ground velocity during passage of earthquake-generated
     seismic waves. Besides, soil amplification effects must be

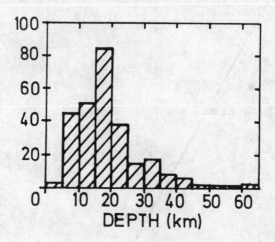

Figure 5.   Focal depth distribution of Fennoscandian earthquakes
            based on macroseismic intensity decay curves. (After
            Ringdal et al [20].)

considered carefully when assessing intensities. Despite its
shortcomings, macroseismic intensity, and in particular the
maximum intensity I_o of an earthquake, is a very valuable
parameter not only for historical earthquakes (where no instrumental observations exist), but also in supplementing
instrumental observations of present-day earthquakes. For
offshore earthquakes, I_o is not readily available, and must
be inferred from land-based observations.

(b)   Area of perception

The size of the area over which an earthquake has been felt
is tied to the size of the earthquake, its depth of focus
and the local wave attenuation characteristics. The area
of perception is often indicated by using an equivalent
radius R corresponding to the 'average' distance from the
epicenter to the furthest point (in any direction) where
the earthquake was felt. This parameter is especially useful
in the present study since maximum intensities cannot be
properly assessed for earthquakes with offshore epicenters.

(c)   Earthquake magnitude

This parameter is a measure of the size of an earthquake
and its definition is physically tied to the kinetic energy

of a wavetrain in a seismogram record. A general magnitude
M definition is

$$M = \log A/T + F(\Delta, H) + C_{S,R}$$

where A = maximum signal amplitude, T = signal period,
$f(\Delta, H)$ = correction term for epicentral distance and focal
depth and $C_{S,R}$ = calibration term depending on seismograph
location and earthquake source region. There are several
different magnitude scales (e.g., m_b, M_s, M_L) and these are
not always compatible. Obtaining a unified definition of
magnitude is a major research problem in seismology today.

The largest known earthquakes in the North Sea and adjacent
waters are shown in Figures 6 and 7. Figure 6 shows all known
earthquakes within the indicated area which have either a magnitude
of at least 6 or a radius of perception of at least 400 km. Figure 7
covers the North Sea, and is an attempt to correlate earthquakes
in the area with known tectonic features. Most of the earthquakes
of the two figures are listed in Table 2.

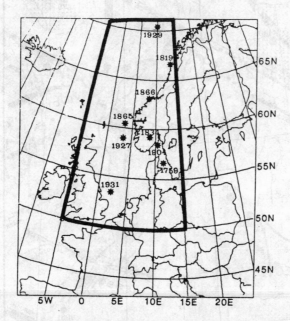

Figure 6.   Map showing the estimated location of the largest known
            earthquakes in the North Sea and adjacent areas. All
            earthquakes shown have been felt over an area of radius
            at least 400 km.

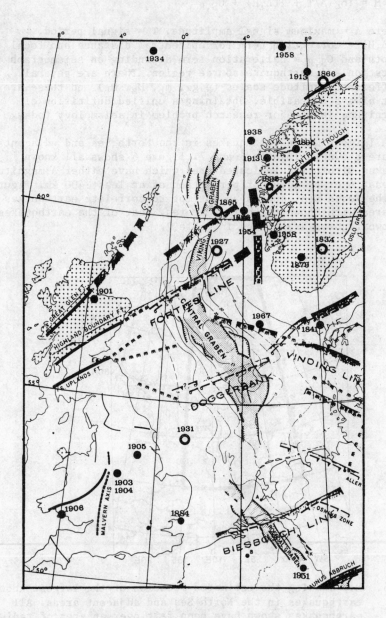

Figure 7.   Tectonic map of the North Sea, with locations of
            some of the most significant earthquakes shown.

	DATE		EPICENTER		DEPTH*	MAGNITUDE	INTENSITY	FELT RADIUS
Year	Day	Hour	Lat	Long	(km)		MM-Scale	(km)
1759	22/12	01.00	57.0N	11.5E	–	6	VII	500
1819	31/8	15.00	66.5N	13.0E	–	6	VII	500
1834	3/9	19.15	59.3N	8.5E	–	5½	VI	400
1841	3/4	02.16	57.0N	8.0E	–	5½	VII+	250
1865	7/5	13.21	60.5N	3.0E	–	5½	VI	400
1866	9/3	01.15	63.2N	9.0E	–	6	VII	400
1879	25/9	00.15	59.0N	7.0E	–	4½	IV	250
1884	22/4	09.15	51.9N	0.9E	–	5½	VIII	150
1886	24/10	23.30	60.5N	4.0E	–	4½	IV	300
1892	15/5	14.51	60.9N	6.0E	–	5½	VII+	240
1895	4/2	23.40	61.9N	7.0E	–	5	VI+	250
1901	18/9	01.24	57.4N	4.1W	30	5.8	VIII	240
1903	24/3	13.30	53.0N	1.7W	25	5.5	VII+	150
1904	3/7	15.21	53.0N	1.7W	20	5.2	VII	142
1904	23/10	10.26	59.2N	10.5E	55	6.4 (2)	VIII	560
1905	23/4	01.37	53.5N	0.9W	40	5.4	VII	119
1906	27/6	09.45	51.6N	4.0W	30	5.3 (3)	VIII	237
1907	14/1	13.03	65.5N	11.0E	–	5	V+	250
1907	27/1	04.58	65.5N	11.0E	–	5½	VI	315
1913	19/7	15.50	64.0N	8.0E	–	5.0 (1)		200
1913	4/8	07.38	61.4N	5.8E	–	4.9 (1)	V+	270
1924	25/7	19.36	72.5N	16.0E	–	5.6 (1)		
1927	24/1	05.18	59.0N	3.0E	–	5.1 (10)	VI	440
1929	10/6	23.03	71.0N	10.0E	–	6.0 (26)		600
1931	7/6	00.25	54.0N	1.4E	(70)	6.0 (27)	(VIII)	600
1934	20/5	19.04	64.5N	2.0W	–	5.6 (1)		
1935	17/7	00.04	65.5N	11.5E	–	5.6 (1)		
1938	11/3	16.08	61.9N	4.2E	–	4.8 (1)	V+	268
1946	11/5	16.25	66.0N	0.5W	–	5.0 (2)		
1946	11/5	18.39	66.0N	0.5W	–	5.1 (8)		
1954	7/7	00.48	59.8N	4.8E	–	4.9 (1)	VI+	350
1955	3/6	11.39	61.9N	4.0E	–	5.0 (2)		
1958	23/1	13.35	65.0N	6.5E	–	5.5 (1)		
1958	6/8	17.16	59.5N	6.0E	–	5.1	VI	300
1959	29/1	23.24	70.9N	7.3E	–	5.8		
1962	15/12	03.48	67.5N	14.2E	–	5	VI	235
1967	21/8	13.41	57.0N	4.9E	–	5.2 (9)		
1975	20/1	10.47	71.7N	14.2E	–	5.0		
1977	6/4	19.31	61.6N	2.5E	–	5.0		

* Depth estimates are those of Karnik (1969).

Table 2. Known, large earthquakes in the North Sea and adjacent
areas. The events listed have either a magnitude of
at least 5.0 or a radius of perception of at least
250 km. Magnitudes are estimated from instrumental
observations when possible (number of stations in
parentheses), otherwise the magnitude values have been
inferred from macroseismic data.

It is noteworthy from Figure 6 that the major occurrence of
large earthquakes takes place along the coast of Norway. From
Figure 7 there seems to be a slight correspondence between the
earthquakes and the major faults of the map, but it is difficult
to give definite conclusions because of the considerable location
uncertainty of the earthquakes in question. This uncertainty is
probably on the order of 25-300 km, highest for offshore and coastal
epicenters and for earthquakes occurring before 1900.

Three of the most significant earthquakes relevant to the
North Sea area are discussed in more detail in the following.

<u>23 October 1904</u>, 59.2°N, 10.5° E, M = 6.4(2), I_o = VIII, R = 560

This earthquake, which had its epicenter in the outer Oslofjord,
is the largest one to have occurred in Northwestern Europe in
modern times, and consequently plays a major role in any seismic
risk analysis for installations in this area. In the Oslofjord
region, hundreds of chimneys collapsed, and tendencies to panic
among the population were observed. The earthquake did not, however,
result in any casualties, and in only few instances were buildings
severely damaged. The earthquake occurred along the Oslo Rift
Zone, which is part of the fracture system in the North Sea. Isoseismals of this earthquake are shown in Figure 8, and it is

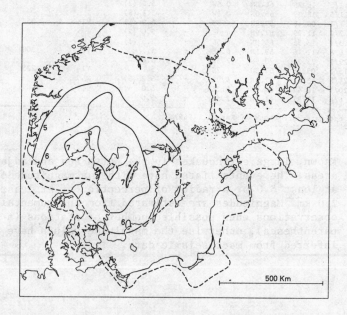

Figure 8.   Isoseismals for the 1904 Oslofjord earthquake.
            (After Båth [17].)

seen that the earthquake was felt over all of Southern Scandinavia
and as far east as the Baltic states. The instrumental measurements
are generally poor for this earthquake, and the magnitude value
of M = 6.4 above is therefore uncertain. It appears that the
true magnitude of this event should be somewhere in the range
of 6.0-6.5 and possibly slightly larger. The depth estimate of
55 km given by Karnik [13] appears too high; a depth of 35 km was
estimated by Austegard [23] based upon known macroseismic data.
A catalogue of macroseismic effects from this earthquake has
been compiled by Kolderup [24].

24 January 1927, 59.0°N, 3.0°E, M = 5.1(10), I_o = VI, R = 440

This is the best documented, large earthquake among those with
an epicenter off the west coast of Norway. Nonetheless, its precise location is uncertain, as is of course the maximum intensity.
The earthquake was felt with intensity V over most of the Norwegian
west coast, and was given a magnitude M = 5.7 by Båth [25] based
upon macroseismic evidence. The instrumental magnitude of 5.1
appears too low in view of the great area over which the earthquake was felt, including most of Scotland, East England, the
Shetland Islands and Norway south of Trondheim. It was also
reported felt in the western part of Denmark, but not in Sweden.
Associated with this earthquake were rumbling or roaring sounds
of low pitch, and in some cases explosion-like bangs. The maximum
intensity of 6 was reached in northeast Scotland, where ornaments
and plaster were thrown down and cracks opened in concrete walls.
Isoseismals for this earthquake are shown in Figure 9. Further
description of macroseismic and instrumental data can be found
in Kolderup and Krumbach [26], Lee [27] and Tyrrell [28].

7 June 1931, 54.0°N, 1.4°E, M = 6.0(27), (I_o = VIII), R = 600

This is the largest known earthquake in the North Sea, and had
its epicenter in the Doggerbank area. From Figure 6 it appears
difficult to associate it to any particular surface fault structure; the uncertainty in the epicenter location should, however,
be kept in mind. Karnik [13] assigns a tentative depth of 70 km
to this earthquake; this value must, however, also be considered
highly uncertain and appears unreasonably deep. The Doggerbank
earthquake was felt in Great Britain, Belgium, the Netherlands,
northern France and Germany; in Norway its effects were mainly
confined to the southwestern coast. The earthquake was in addition felt in Denmark, as far east as Copenhagen, according
to Lehman [8].

The earthquake damaged buildings in a coastal belt of Great
Britain, some 30 km wide, extending between Scarborough and
Grimsby, displaced boulders from Castle Hill, Scarborough;

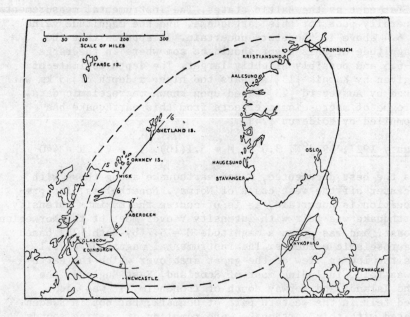

Figure 9. Isoseismals for the 1927 North Sea earthquake.
(After Tyrrell [28].)

enlarged cracks in Lincoln Cathedral and changed the water levels
of four wells in the area, among many other effects. People in
boats passing through the epicentral region at the time heard
sound of submarine origin which resembled muffled explosions and
noticed a heavy, confused swell which developed in a calm sea
but must have subsided relatively rapidly since no tsunamis were
seen on the adjacent coasts of England.

Isoseismals for this earthquakes are shown in Figure 10.
For further description of the 1931 earthquake, reference is
made to Versey [29].

There are also several other earthquakes (with onshore and
coastal epicenters) which may be significant in assessing seismic
risk in the North Sea. In the British Isles, earthquake swarms
have been recorded in Comrie, Scotland, and otherwise several
strong shocks (intensity VIII-IX) have been reported in the past
200 years. The most significant appears to be the 22 April 1884
earthquake near Colchester, which caused considerable damage to
buildings in the area (Karnik [14]). In the Netherlands-Belgium
region, the largest known event appears to have been the earthquake of March 14, 1951, near Enskirchen (I_o = VIII, M = 5.4).

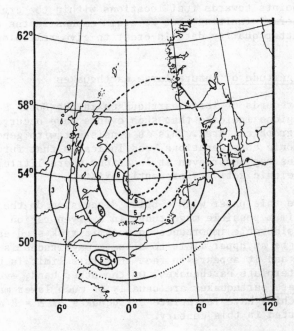

Figure 10.   Isoseismals for the 1931 Doggerbank earthquake.
             (After Versey [29].)

DISCUSSION

Precision of estimated earthquake parameters

Obviously, macroseismic data are not suitable in the North Sea
area, because of lack of seaward observations. Until recently,
instrumental data are neither well suited for precise locations
due to relatively few station recordings and poor azimuthal
coverage. For example, different seismological agencies have
reported epicenter locations for the very same earthquake with
separations exceeding 100 km! Nevertheless, instrumental data
are signifcantly better than macroseismic for epicenter locations in coastal areas, as well as offshore.

     Problems involved in properly estimating earthquake in-
tensity (macroseismic) or magnitude (instrumental) are similar
to those involved for a proper epicenter location, i.e., a general
lack of sufficient observational data. Furthermore, a useful,
local magnitude scale has not been established for the general
North Sea area. Precise focal depth estimates are difficult

even where extensive observational data are available. All available evidence points towards foci locations within the crust and so-called normal faulting, but we again emphasize the need for more and better quality data in order to give reliable conclusions.

Location and magnitude of future large earthquakes

The historical records of large earthquakes in the North Sea and adjacent regions indicate that such events have occurred over a wide geographical area. This is consistent with general intraplate tectonic considerations, and indicates that future, large earthquakes may occur even at locations where little indication of seismic activity presently exists.

We conclude this paper with a brief discussion on the problem of 'maximum possible magnitude' in a given region, which is of considerable importance in seismic risk studies. That there must be an upper limit to seismic magnitudes is physically obvious, and it appears in fact that this limit is between 9 and 10 for interplate earthquakes. On the other hand, even the largest intraplate earthquakes are usually of much lower magnitudes; only in China have intraplate earthquakes of M = 8 and above been reported in this century.

Chinese earthquakes do appear to be anomalous in this respect, and it would probably be unduly conservative to assume that such large earthquakes can be expected in any intraplate area. With the limited data that are available for the North Sea area, it is still necessary to consider other intraplate areas in order to get indications as to the maximum magnitude that might be expected. The region which points itself out in this respect is the eastern part of North America. Like the North Sea, this region is well removed from any plate boundary, and tectonically the Appalachian mountain chain, which is the dominant geomorphological lineament of the region was formed during the same orogeny as the Caledonides when America and Eurasia 'collided'. Post-glacial uplift is also a common feature to parts of eastern North America and Northwest Europe.

Several large earthquakes of magnitude 7-7.5 have struck eastern North America during the past few hundred years. They included the 1663 St. Lawrence Valley earthquake, and three major earthquakes of New Madrid, Missouri, during the winter 1811-1812, the 1886 Charleston, South Carolina, earthquake and the 1929 earthquake near the Grand Banks of Newfoundland. Even though such earthquakes are rare, it is nonetheless firmly established that earthquakes of magnitude 7 and slightly above are possible in widely distributed locations in eastern North America (Stein et al [30]). Other intraplate areas from which

we have reliable records of earthquakes of similar magnitudes
include the USSR and Australia.

In conclusion, while the largest known earthquakes in the
North Sea and adjacent regions have been of magnitude 6.0-6.5,
we cannot infer that the maximum possible magnitude in this
area would be so limited. With the present knowledge, it appears
that a seismic risk analysis for the North Sea area must take
into account the (admittedly low) probability of future occurrence of earthquakes in the magnitude range 7.0-7.5 in this
region.

REFERENCES

[1]   Doornbos, D.J., 1981: Seismic moment tensors and kinematic
       source parameters. Geophys. J. R. astr. Soc., 69, 235-251.

[2]   Knopoff, L., 1981: The nature of the earthquake source, in:
       E.S. Husebye and S. Mykkeltveit (eds.), Identification of
       Seismic Sources - Earthquake or Underground Explosion,
       D. Reidel Publ. Co., Dordrecht, the Netherlands, 49-70

[3]   Madariaga, R., 1981: Dynamics of seismic sources, in:
       E.S. Husebye and S. Mykkeltveit (eds.), Identification
       of Seismic Sources - Earthquake or Underground Explosion,
       D. Reidel Publ. Co., Dordrecht, the Netherlands, 71-96.

[4]   Richter, C.F., 1958: Elementary Seismology, W.H. Freeman
       & Co., San Francisco, Calif. USA, 768 pp.

[5]   Kanamori, H., 1977: The energy release in great earthquakes,
       J. Geophys. Res., 82, 2981-2987.

[6]   Keilhau, B.M., 1836: Efterretninger om Jordskjælv i Norge,
       Magasin for Naturvidenskaperne, Bd 12, 83-165.

[7]   Kolderup, C.F., 1913: Norges jordskjælv, Bergens Museums
       Aarbok 1913, 8, 152 pp.

[8]   Lehman, J., 1956: Danske jordskjælv, Bull. Geol. Soc.
       Danmark, 13, 88-103, Copenhagen, Denmark.

[9]   Davison, C., 1924: A history of British earthquakes, Cambridge,
       416 pp.

[10]   Dollar, A.T.J., 1950: Catalogue of Scottish earthquakes 1916-
       1949, Univ. of Glasgow.

[11] Ritsema, A.R., 1966: Note on the seismicity of the Netherlands,
      Proc. Koninkl. Nederl. Ak. Wetensch. 69, No. 2, 235-239.

[12] van Gils, J.M., 1956: La seismicite de la Belgique, Acad.
      Royale Belgique, Com. Nat. Geogr. Atals de Belgique, Plandr. 10
      11 pp.

[13] Karnik, V., 1969: Seismicity of the European Area, Vol. I,
      D. Reidel Publ. Co., Dordrecht, 364 pp.

[14] Karnik, V., 1971: Seismicity of the European Area, Vol. II,
      D. Reidel Publ. Co., Dordrecht, 218 pp.

[15] Horrebow, C., 1765: Beretning om Jordskiaelvet, som skeede
      d. 22 dec. Ao. 1759, Kongelig Dansk Videnskapsselskab
      Skrifter, Bd 9, 261-372, Copenhagen, Denmark.

[16] Wood, H.O. and F. Neumann, 1931: Modified Mercalli scale of
      1931, Bull. Seism. Soc. Am., 21, 277.

[17] Båth, M., 1972: Zum studium der seismizität von Fennoskandia,
      Gerlands Beitr. Geophysik, 81, 213-226.

[18] Ambraseys, N.N., 1971: Value of historical records of
      earthquakes, Nature, 232, 375-379.

[19] Bungum, H., E.S. Husebye and F. Ringdal, 1971: The NORSAR
      array and preliminary results of data analysis, Geophys.
      J.R. astr. Soc., 25, 115-126.

[20] Ringdal, F., E.S. Husebye, H. Bungum, S. Mykkeltveit
      and O.A. Sandvin, 1982: Earthquake Hazard Offshore Norway,
      NTNF/NORSAR, in press.

[21] Bungum, H., B.Kr. Hokland, E.S. Husebye and F. Ringdal,
      1979: An exceptional intraplate earthquake sequence in
      Meløy, Northern Norway, Nature, 280, 32-35.

[22] Bungum, H. and J. Fyen, 1979: Hypocentral distribution,
      focal mechanisms and tectonic implications of Fennoscandian
      earthquakes, 1954-1978. Geol. För. Stockh. Förhl., 101, 261-
      271.

[23] Austegaard, A., 1975: The 1904-earthquake in the Oslofjord
      area, Appendix C of the report: Phase 1 studies - Preliminary
      seismic design criteria. Five alternative sites around the
      Oslofjord. Report for NVE-statskraftverkene, Oslo, Norway.
      Prepared by Dames & Moore, New Jersey, 1975.

[24] Kolderup, C.F., 1905: Jordskjælvet den 23. oktober 1904, Bergens Museums Aarbok 1905, 1, 1972 pp.

[25] Båth, M., 1956: An earthquake catalogue for Fennoscandia for the years 1891-1950, Sveriges Geol. Unders. Årbok 50, No. 1, 52 pp, Stockholm, Sweden.

[26] Kolderup, N.H. and G. Krumbach, 1931: Das Nordseebeben von 24 Januar 1927, Zeitschr. für Geophysik, VII, 225-232

[27] Lee, A.W., 1932: The North Sea earthquake of 1927 January 24, M.N.R.A.S. Geophys. Suppl. III, 21-30.

[28] Tyrrell, G.W., 1932: The North Sea earthquake Jan 24, 1927, Trans. Geol. Soc. Glasgow XIX, 10-29.

[29] Versey, H.C., 1939: The North Sea earthquake of 1931 June 7, M.N.R.A.S. Geophys. Suppl. IV, No. 6, 416-423.

[30] Stein, S., N.H. Sleep, R.J. Geller, S.C. Wang and G.C. Kroeger, 1979: Earthquakes along the passive margin of Eastern Canada, Geophys. Res. Lett., 537-540.

SOME PROBLEMS CONCERNING THE SEISMICITY OF THE NORTH SEA AREA

J.-M. Van Gils

Observatoire Royal de Belgique

Abstract.

Seismicity is considered from a double point of view : its
"static" aspect that leads to the seismotectonic conformation
of the area, while its dynamic concept aims to the evaluation
of earthquake hazard. After having examined the case of the North
Sea it is stated that an isoseismal map of the area is still missing and should be undertaken. The occurrence of seismic seawaves
in the North Sea has an extremely low probability. Conclusions
and recommendations.

1.- Seismicity and Seismic activity.

Seismicity and seismic risk being the main subjects of the
workshop, some consideration should be made as regards the use
of these terms. Principally, as the first one can be understood
in two different ways, its concept must be made clear.

From the static point of view, seismicity means the set of
events affecting the area under investigation, while from the
dynamical viewpoint it must be consideredin its space-time context. In the first case, seismicity answers the question : is the
area concerned affected by earthquakes or not, whereas the second
concept replies to the double question : where and when do seismic events strike the area.

In the latter cas, the term "seismicity" should be replaced
by "seismic activity". The first acceptation of seismicity obviously leads to drawing up a seismotectonic map of the area, while
seismic activity aims to the evaluation of earthquake hazard, one

77

A. R. Ritsema and A. Gürpinar (eds.), Seismicity and Seismic Risk in the Offshore North Sea Area, 77–88.
Copyright © 1983 by D. Reidel Publishing Company.

of the main factors in "seismic risk" assessments.

2.- Static Seismicity.

Coming back to the statical aspect of seismicity, the only
aspect that will be considered further on, an area may be struck
by earthquakes having their epicentre either inside or outside
the area itself.

As regards the North Sea, its seismicity is implied by its
own epicentres and those located in the surrounding countries.
This is illustrated in Fig. 1, extracted from (1) ; and reveals
that the seismicity of the North Sea is rather on the lower level
in comparison with that of the encircling region. As a result,
the area under investigation doesn't act as an independent seismic unit. More, its borders may be affected by the adjacent zones
having a relatively higher seismicity level. That means that the
coastal belt should be given special consideration.

These reciprocal effects of earthquakes with epicentres both
onland and offshore are demonstrated in fugures 2 up to 8.

Figures 2, 3 and 4 show the influence of the three major
earthquakes having occurred in the North Sea :
a) the Oslofjord earthquake of October 23d 1904 ;
b) the North Sea earthquake of January 24th 1927, and
c) the Doggerbank earthquake of June 7th 1931.

On the contrary, figures 5, 6, 7 and 8 illustrate the effects
of on land earthquakes on the coastal zone and may suggest an
eventual outline of extrapolated isoseismals on the sea :
a) the Düren earthquake of April 4th 1640 ;
b) the Tirlemont earthquake of September 18th 1692 ;
c) the Düren earthquake of December 26-27th 1755, and
d) the Brabant Massive earthquake of June 11th 1938.

3.- The isoseismal map of the North Sea.

Thus far, only the effects of individual earthquakes have
been examined, but the same procedure is applicable when considering a set of seismic events having occurred in a selected
area during a given period. Doing so, the maximum influence observed at every place of the region can be mapped out. Such a
chart has been drawn for the region encompassing the territories
of Belgium, Luxemburg, the Netherlands and the western part of
Germany.

It is based on data set grouping all epicentres with an
intensity I V of the MSK-Intensity Scale 1964 for a period running from 1500 AD up to 1975. It reflects the distribution of

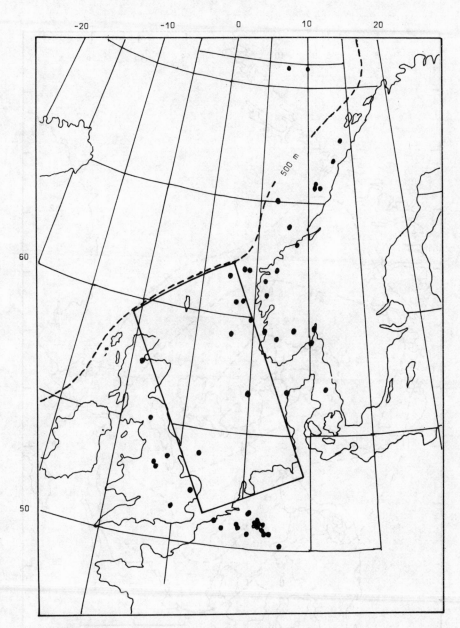

FIGURE 1 - The epicenter map with delineation of the area.

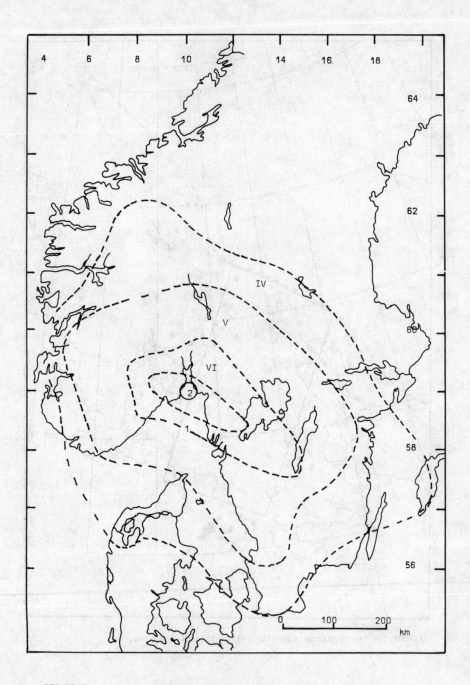

FIGURE 2 -   The Oslofjord earthquake of 23 October 1904.

the maximum observed intensities and the isolines, after some
smoothing, were traced out taking into account the main features
of the regional tectonics.

Although the map shows where the isolines are crossing the
coast-line, nothing was tried to draw them out over the sea.
Nevertheless, such a map may serve as an example to be followed
by all the other countries bordering the North Sea in order to
investigate how far in sea the isolines could be extended and,
eventually, join the isoseismals resulting from other sets of
earthquakes. (see Fig. 9).

4.- Seismic Sea-waves.

Another point, noteworthy for consideration, consists in the
fact that seismic sea-waves might be generated in the North Sea
area although tsunamis are relatively infrequent in the Atlantic
region, so much the less in the region under investigation.

As Chairman of the special Sub-Commission on Tsunamis of the
European Seismological Commission, N.N. Ambraseys prepared the
first Progress Report concerning a systematic study of the seismic sea-waves in the European area. The map he published reveals
the North Sea area as being a blank one, while the main tsunamis
are concentrated in the Mediterranean Basin.

Notwithstanding this, it must be remembered that the Lisbon
earthquake of 1755 produced large sea-waves that entered the
Channel and the North Sea with an amplitude great enough to be
noticed in several harbours. (See Fig. 10).

5.- Conclusions and Recommendations.

   a) - The Data Base.

          The prime thing to stress on is the disparity of the
          data base as it is distributed over several countries,
          the consequence being its non-homogeneity.
          All data relevant to North Sea seismic events, historical
          and present day ones as well, should be collected where-
          after they should be confronted among each other on a
          common base in order to eliminate misinterpretations,
          wrong evaluations, duplicates, a.s.o. ..

          Such an undertaking has already been made between Luxem-
          burg, Germany, the Netherlands and Belgium for the iso-
          seismal map of this part of Europe. For the moment the
          same is on the way for some earthquakes between the Im-
          perial College (London) and Uccle.

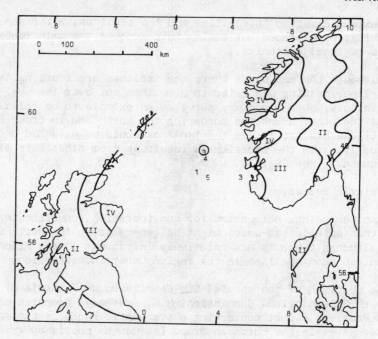

FIGURE 3 - The North Sea earthquake of 24 January 1927.

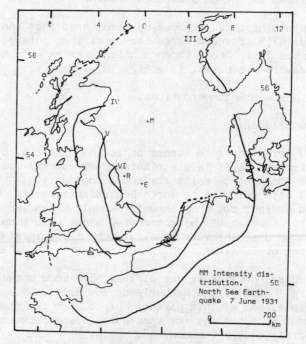

MM Intensity dis-
tribution.
North Sea Earth-
quake 7 June 1931

FIGURE 4 - Doggerbank earthquake of 7 June 1931.

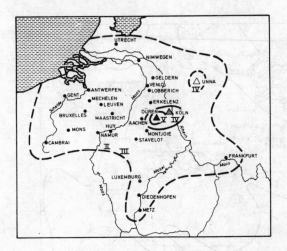

Figure 5. - The Düren earthquake of 4 April 1640.

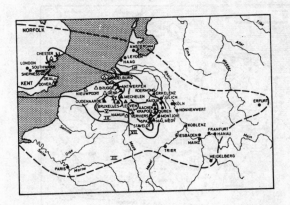

Figure 6. - The Tirlemont earthquake of 18 September 1962.

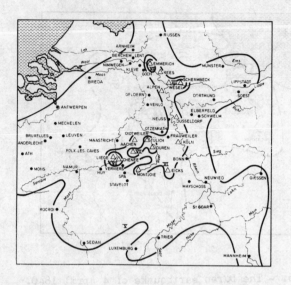

Figure 7. - The Düren earthquake of 26-27 December 1755.

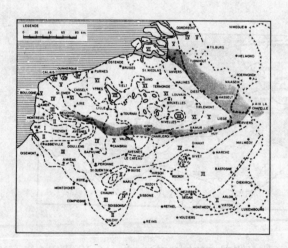

Figure 8. - The Brabant Massive earthquake of 11 June 1938.

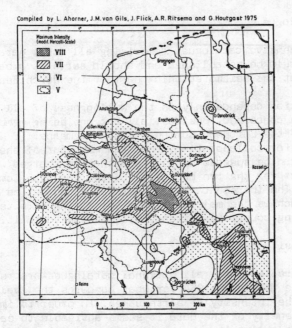

Figure 9. - The Isoseismal Map of western Europe.

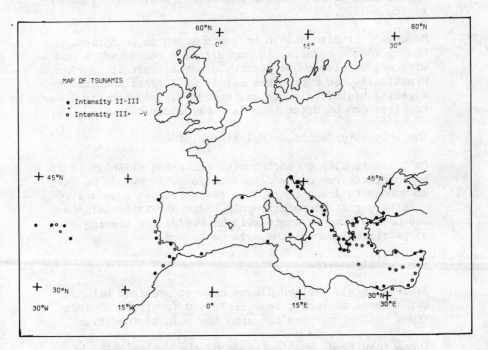

Figure 10. - The Map of Tsunamis in the European Area.

b) - The Catalogue.

   A comprehensive catalogue, comprising all focal parame-
   tres, should be established and should serve as a basis
   to map out the seismotectonic chart of the North Sea.

   Its creation depends only on the willingness of all the
   countries concerned. Besides, all but one being member-
   states of the European Communauties what means, if Norway
   is ready for collaboration, that the make out of the catalogue can be speeded up as the author is charged with
   the object for the ten countries of the European Communauties. The increment to that catalogue can not be extended so much as the major earthquakes, having affected the
   surrounding countries, already will be listed in it.

c) - The Intensity Scale.

   For the greater part, all national catalogues are referred to the MSK-Scale 1964 for what concerns the evaluation of the intensity of earthquakes. In order to improve
   the homogeneity of the data it seems advisable to each
   country to revise its national catalogue making reference
   to the last version of the MSK-Scale, i.e. MSK - Intensity Scale 1980.

   Moreover, it also should be recommended to countries,
   having common borders, to consult each other before drawing any conclusion as regards eventual seismic problems.
   Practically, such problems mainly will arise when investigating historical data as, nowadays, the epicentral pa-
   rametres can be determined far more accurately.

d) - The Intensity-Magnitude Relationship.

   As the intensity of earthquakes occurring offshore is to
   be inferred from on-land observations, it should be recommended to investigate a to the North Sea adequate intensity-magnitude relationship. Such a relationship based on instrumental data will be useful for tracing out
   an isoseismal map of the area concerned.

e) - The Questionnaire.

   As long as all dwelled places are not provided with a
   seismoscope at least, near-field and far-field observations must be carried out with the help of inquiries.

   These inquiries, when not made by the seismologist himself across the struck area, are mainly based on questi-

onnaires disseminated through the region or on observations
made by the population and transmitted to the seismological
centre by different ways.

As far as this concerns, very few or even none seismological
observatories are using the questionnaire indicated by the
Handbook on Practice in Seismological Observatories by P.
Willmore, given by W. Sponheuer (1967) and fairly well adapted to the MSK-Scale.
Almost all seismological stations practise their own questionnaire and, doing so, create some hindrance to achieve the needed homogeneity. Therefore, it should be agreed, for general
use, on a unique questionnaire including all items necessary
to produce trustworthy, homogeneous macroseismic maps in all
parts of the region under investigation.

General agreement should be looked for also regarding the
ways and to who the questionnaire is to be addressed.

After all, the recommendations are :

- an improved, homogeneous data base ;
- a trustworthy catalogue ;
- a generally agreed intensity scale ;
- an applicable intensity-magnitude relationship ;
- a comprehensive questionnaire, and

as a result of all this :

- a seismotectonic chart, and
- an isoseismal map of the North Sea Area.

REFERENCES.

1.- A.R. RITSEMA : Comments on the OVE ARUP REPORT :"Earthquake
                  Effects on Platforms and Pipelines in the U.K.
                  Offshore Area" - K.N.M.I. - De Bilt - January
                  1981.

2.- N.N. AMBRASEYS : A personal communication on the Oslofjord
                     Earthquake of 23 October 1904.

3.- N.N. AMBRASEYS : A personal communication on the North Sea
                     Earthquake of 24 January 1927.

4.- M. KIELMAS : A Re-appraisal of available seismological Data
                 for the NOrth Sea. (1975).

5.- A. SIEBERG : Beitrage zum Erdbebenkatalog Deutschlands und
                 angrenzender Gebiete für die Jahre 58 bis 1799.-
                 Berlin - 1940.

6.- W. SPONHEUER : Erdbebenkatalog Deutschlands und angrenzender
                   Gebiete für die Jahre 1800 bis 1899. -
                   Berlin 1952.

7.- O. SOMVILLE : Le Tremblement de Terre belge du 11 juin 1938.-
                  Duculot - 1939.

8.- L. AHORNER, J.-M. VAN GILS, J. FLICK, A.R. RITSEMA and
    G. HOUTGAST : Seismic Zoning Map of Northwestern Germany,
                  Belgium, Luxemburg and the Netherlands. - 1975.

9.- J.-M. VAN GILS and Y. ZACZEK : La Séismicité de la Belgique et
                  son Application en Génie paraséismique. -
                  Annales des Travaux Publics de Belgique - 1978/6.

10.- V. KARNIK : Seismicity of the European Area / 2 -
                 Reidel - Dordrecht 1971.

NON-RANDOM DISTRIBUTION OF EARTHQUAKES IN THE NORTH SEA BASIN AREA

A. Reinier Ritsema

Royal Netherlands Meteorological Institute,
De Bilt, The Netherlands

ABSTRACT

Seismicity and seismic hazard are considered for different
sectors of the North Sea basin. The NE sector (Norway) seems to
be 3 times more hazardous than the W sector (UK) and 5 times
more than the SE sector (Netherlands). Also inside the three
sectors important differences occur. Risk evaluation, therefore,
should be made for each sector or sub-sector separately.

The maximum credible earthquake in the region is found to be
of the order of magnitude $6\frac{1}{2}$. Results for the Eastern USA should
not be used indiscriminately for W. European shelf regions because
of vast differences in activity between the areas during the past
300 years.

SEISMICITY OF THE NORTH SEA REGION

The epicenterplot of the NGSDC earthquake data file for the
North Sea basin and adjacent land regions is shown in the figure 1.
Earthquakes in the magnitude range 3-6 are present for the period
1926-1980. It is known that the NGSDC file is incomplete for the
early- and pre-instrumental period and for the smaller magnitudes.
Nevertheless, the general lines of seismic activity in the region
are correct. For more complete data lists for the region reference
is made to Båth 1954, 1956, 1972, Husebuy et al. 1978 for the
Norwegian sector; to Lilwall 1976, Burton 1978 for the UK sector;
and to Ritsema 1966, Ahorner et al. 1976, Van Gils et al. 1978
for the Netherlands sector.

89

A. R. Ritsema and A. Gürpınar (eds.), Seismicity and Seismic Risk in the Offshore North Sea Area, 89–99.
Copyright © 1983 by D. Reidel Publishing Company.

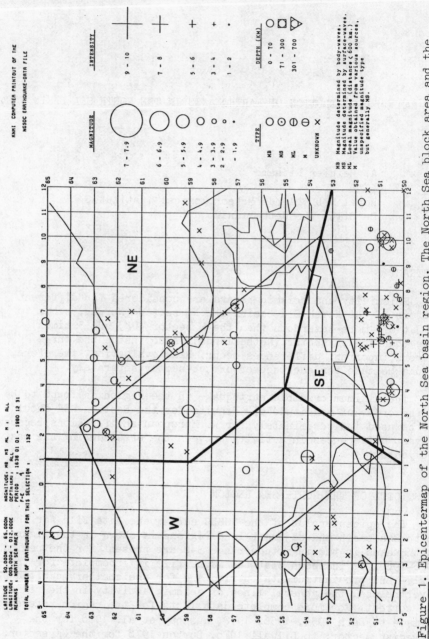

Figure 1. Epicentermap of the North Sea basin region. The North Sea block area and the division of the region in W, NE and SE sectors are indicated.

The frequency distribution of earthquakes in the region is
given in the figure 2, based upon 100 years of observations in the
period 1878-1978 as given in Ritsema, 1980 and 1981. The greater
block comprises the total area of figure 1 with an extension of
6° in the North and 3° in the East direction; the limitations of
the North Sea block proper are as indicated in the figure 1.

On the basis of relative numbers of shocks, separate distributions have been calculated for the W, NE and SE sectors of the
region, assuming that an equal b-value is valid for the frequency-
magnitude curves for the sub-sectors (figure 3). For the W and NE
sectors the frequency distribution is found to be very similar,

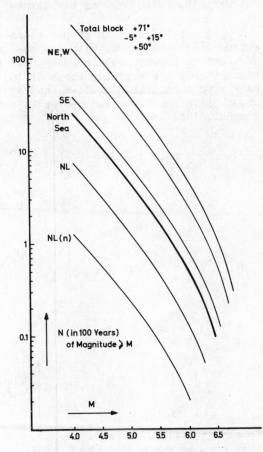

Figure 3.
Scetch of the relative
numbers of earthquakes in
sub-sectors of the North
Sea basin region.

both being about 4 times that of the SE sector and 6 times that of
the North Sea block proper. Note that these numbers do not give a
direct indication as to the relative seismicity of the sectors
since the areal surface of the sector regions has not been taken
into account here.

Obviously, there are three main concentrations of epicenters in the region:

\-   The Western sector with the UK including the western part of the North Sea shelf. The probably strongest earthquake in the area was that of Colchester, 1884, April 22 of magnitude $5\frac{1}{2}$-$5\frac{3}{4}$, or that of the Doggersbank, 1931, June 7 with a magnitude of $5\frac{3}{4}$-6.

\-   The NE sector with S. Norway and Denmark, including the NE part of the shelf region. The strongest event was that of Oslo, 1904, October 23 with a magnitude of about $6\frac{1}{4}$.

\-   The SE sector with the Netherlands, Belgium and NW Germany, including the SE part of the North Sea basin with the magnitude 6 earthquakes of 1692, September 18, Tirlemont, Belgium, and 1756, February 18, Düren, FRG as the largest events.

The North Sea basin proper seems to be less active than the surrounding landmasses, with the one exception of the NE sector adjacent to Norway. The largest known events located within the basin itself are probably those of 1931, June 7 at the Doggersbank with a magnitude of $5\frac{3}{4}$-6, and of 1927, January 24, magnitude about $5\frac{1}{2}$ in the Norwegian sector (see also Ambraseys, 1982 and Ringdal, 1982).

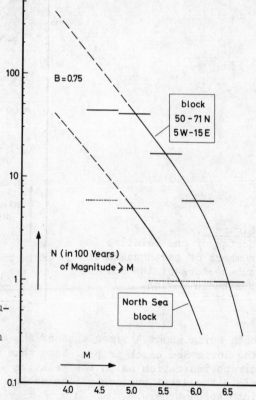

Figure 2.
Frequency-magnitude distribution for earthquakes of the years 1878-1978 in the North Sea area s.l. and s.s.

SEISMICITY OF THE NETHERLANDS

    In figure 4 the epicenters in the Netherlands and adjacent
regions are indicated. The locations show a clear relation with
the Graben structure in the SE part of the country, constituting
the NW extension of the Lower Rhine or Rur Graben.

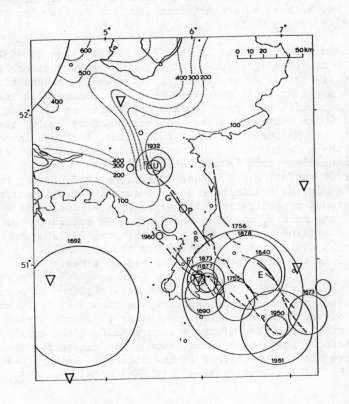

Figure 4.
Seismicity of the Netherlands and adjacent
regions (from Ritsema 1966). Earthquakes are
shown as circles with the appropriate year.
The areal surface of the circles is proportional to the root of the energy or the
seismic moment of the earthquake. The main
faultlines in the underground and contour-
lines for the base Pleistocene are indicated.
Seismic stations (in 1966) are given as
triangles.

Since 1500 we have found 152 epicenters on Netherlands territory (see also Van Rummelen, 1945). Six earthquakes were followed
by greater or smaller numbers of aftershocks, not included in the
list of 152 epicenters. 53% of the total of 152 took place in the
Southernmost part of the country, the province of S. Limburg,
29% in the N. Brabant province just North and Northwest of it,
and 18% in the Northern half of the country, North of the E-W
rivers in the Netherlands. Apart from the numbers also the size of
the shocks decreases Northward.

The main historic event in the Netherlands, that of 1932,
November 20 caused slight damage to ill-built constructions in
Uden (U in figure 4). By re-levelling measurements recent and
sub-recent dip-slip fault movements have been observed in the
field up to 1 cm/20 year near Roermond (R in figure 4). Near
Gemert (G, 40 km Northward along the Peel boundary fault) the
amount has been measured as 0.6 of this value. Although in the
Southeast the main boundary fault can be observed in the field
as an inconspicious step in the landscape where at several places
deep-water is surfacing, no sudden displacements following recent
small shocks in the region have been measured. A fault motion in
the underground probably finds its expression at the earth's
surface in the form of a continuous creep. The subsidence at the
North Sea coastline is about 600 m for the base Pleistocene,
which gives a similar rate of motion for the past million year
as found along the main fault line for recent times.

From these available historic data the frequency of occurrence
of earthquakes on Netherlands territory is about six times smaller
than that of the total SE sector, including the North German and
Belgian earthquakes. That of the Northern half of the country is
another factor 5 smaller (Ritsema, 1966). The curves for these
regions have also been scetched in the figure 3, using a b-value
equal to that of the total material (NL and NL(n) respectively).

SEISMICITY OF THE OFFSHORE NORTH SEA AREA

In the group of 152 Netherlands earthquakes since 1500 there
are only three which possibly had their epicenter within the
offshore North Sea basin:
-     1883, March 17: felt in the region Alkmaar, The Hague, Utrecht,
especially in the Haarlem-Leiden region; radius of perceptibility
of the order of at least 70 km; possible epicenter 52.4°N 4.2°E,
magnitude around 4.5 ?
-     1850, September 9: felt in the coastal region Bloemendaal-
Hillegom, some cracks in chimneys were reported; possible epicenter
around 52.3°N 4.3°E, magnitude 4-4.5 ? Aftershock on December 19,
same year.
-     1833, December 2: felt in Haarlem, 20 seconds duration, somewhat more distant than the other two ?

The scarce data available for the Netherlands shelf region
suggest that the activity of this part of the North Sea basin is
more like that of the Northern half of the Netherlands than that
of the average SE sector, and that it is definitly less than in
the S. Limburg province and adjacent NW German region. We assume
a shelf area five times greater than that of the Northern part of
the country and extrapolate the low on-land seismicity into the
sea-covered area. The activity in number of shocks of this part of
the North Sea then is conservatively estimated to be six times
smaller than that of the SE sector in total.

For the UK sector of the shelf the activity is hard to evaluate.
Only a few more heavy events have been located in the area. In a
200-years period (see also Lilwall, 1976) there were only 3 offshore earthquakes of magnitude $5+\frac{1}{4}$ and higher, against 10 onshore. Since shocks of this size in the period covered will have
been adequately observed on land also if the epicenter was far
on the UK shelf region of the North Sea, we may use this factor
as a base for extrapolation of land seismicity in offshore seismicity. We than conclude that in comparison with the total W sector,
including the onshore UK, the UK North Sea basin activity is of
the order of 1 in 4 less.

For the Norwegian sector, in contradiction to the former two
sectors, the seismicity for the greater part is concentrated offshore with only 20-30% on the mainland. This means that the
activity of the Norwegian shelf part of the basin is about 75%
of that of the total NE sector.

From these considerations it is obvious that big differences
exist between the seismic hazards in the different sectors of the
basin. A rough estimate of the relative hazard per unit area in
the basin on the basis of these data, including the relative areal
surface of the Norwegian, United Kingdom and Netherlands sector
taken to be around 7:7:2, is found to be

N : UK : NL = 6 : 2 : 1.2

Obviously, the reliability of these figures depends on the completeness of the available data. In a region of low seismicity such as
this one, it is very difficult indeed to compile a homogeneous
and long-enough data-set. The ratios given here, therefore, should
be considered as such preliminary values that are liable to correction when more complete data will be gathered.

With these restrictions in mind it is possible to construct
the curve of the probability that certain values of acceleration
are exceeded. We use a conversion of the magnitude of the shock
via epicentral intensity and attenuation of seismic intensity with
epicentral distance to acceleration (Ritsema 1980, 1981). The result

for the North Sea basin proper was given as the relation of the
recurrence time in years for given values of acceleration assuming
a random distribution of epicenters in the area (Ritsema 1980).

We have seen now that this supposition is untrue and that
distinct differences exist between sectors of the basin. In the
figure 5, therefore, curves could be drawn not only for the
total North Sea basin block, but also for the Norwegian, the UK
and the Netherlands sector separately. The base of this figure is
from a paper by Costes (1981) in which a compilation is made of
the exceedance probability of certain values of acceleration for
different seismic regions. It is seen that the curves for the
North Sea sectors are in reasonable accordance with data for
regions with similar type of seismicity, such as those for parts
of NW Germany, Scandinavia and France.

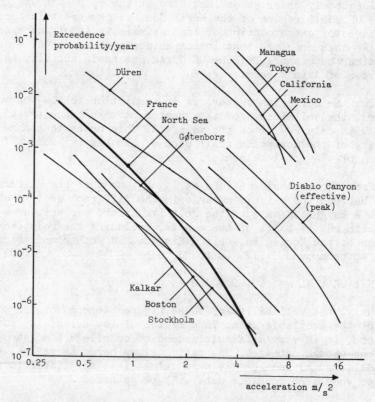

Figure 5. Exceedance probability in years of seismic accelerations for the North Sea basin area.
For comparison, the lines for other seismic regions
as given by Costes (1981) are included.

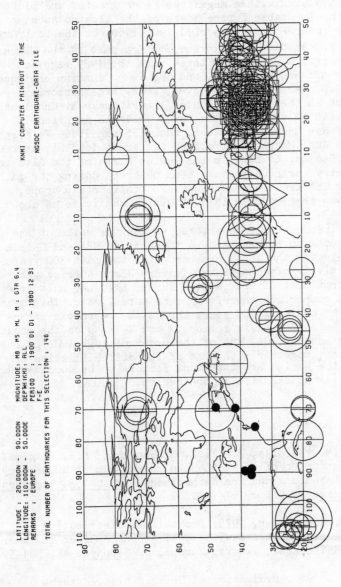

Figure 6. Epicentermap of Europe and part of N. America with all the earthquakes in the period 1900–1980 of magnitude $6\frac{1}{2}$ or greater. The biggest symbols are of M 8 or higher, middle size M 7–7.9 and smallest M 6.5–6.9. For legend see figure 1. Full circles are historical events between 1650 and 1900 of magnitude 7 or higher in continental USA/Canada. In continental Europe no such events did occur.

THE MAXIMUM CREDIBLE EARTHQUAKE

For information on the maximum credible earthquake in the region all earthquakes from the NGSDC file within the region 20-90°N and 105°W 50°E with a magnitude $6\frac{1}{2}$ or greater are plotted in the figure 6. In stable Europe North of the Alps no such earthquake has occurred in the period 1900-1980. On the contrary, several heavy shocks did occur in the Eastern part of the USA and Canada. In the period since 1900 these were the 1929 magnitude 7.2 earthquake off New Foundland, and the 1925 Canadian earthquake of magnitude 7.0 near Quebec. Moreover, some large historical events are known, such as the 1886 magnitude $7\frac{1}{2}$ earthquake at Charleston, S. Carolina (with 60 victims), the three 1811-'12 magnitude 8 earthquakes of New Madrid, Missouri, the 1755 magnitude 7+ earthquake off Cape Ann, Massachusetts, and the great 1663 St. Laurence earthquake in Canada. Nowhere in the region North of the Alps in Europe such heavy earthquakes have been observed during the past 300-500 years. The greatest earthquake in continental Europe, most likely, was that of Oslo in 1904 with a magnitude of about $6\frac{1}{4}$.

It seems inappropriate therefore, to use the value of the maximum earthquake to be expected in the Eastern USA and Canada indiscriminately for the North Sea basin. In the past 300 years the seismic activity of continental Eastern North America was much greater that that of continental Europe also per unit area and any direct comparison therefore seems unrealistic. This means that on the time basis of 100 years the maximum credible earthquake for the North Sea basin region is not $7-7\frac{1}{2}$ as on the coastal regions of the Eastern USA, but of the order of magnitude $6\frac{1}{2}$ only. This observation, together with that of the differentiation of hazard within the basin, are important factors to be considered in studies of earthquake risk in the region.

REFERENCES

Ahorner, L., J.M. van Gils, J. Flick, G. Houtgast, A.R. Ritsema, 1976. First draft of an earthquake zoning map of NW Germany, Belgium, Luxemburg and the Netherlands, Publ. 153, Kon. Ned. Meteor. Inst. De Bilt: 167-170.

Ahorner, L., M. Rosenhauer, 1975. Probability distribution of earthquake accelerations with application to sites in the Northern Rhine area, Central Europe, J. Geophys. 41: 581-594.

Ambraseys, N.N., 1982. Evaluation of Seismic Risk, Present Volume:

Båth, M.,1954. Seismicity of Fennoscandia and related problems, Gerlands Beitr. z. Geophysik 63: 173-208.

Båth, M., 1956. An earthquake catalogue for Fennoscandia for the years 1891-1950, Sveriges Geol. Unders Arsbok 50: pp 52.

Båth, M., 1972. Zum Studium der Seismizität von Fennoskandia, Gerlands Beitr. z. Geophysik 81: 213-226.

Burton, P.W., 1978. Perceptible earthquakes in the United Kingdom, Geophys. Jour. Royal Astr. Soc. 54: 475-480.

Costes, D., 1981. Probabilistic approach of reference seismic ground motions, CSNI Report no. 44, Proceedings Volume III: 82-94.

Husebuy, E.S., H. Bungum, J. Fyen, H. Gjoystdal, 1978. Earthquake activity in Fennoscandia between 1497 and 1975 and intraplate tectonics, Norsk Geol. Tidsskrift 58: 51-68.

Lilwall, R.C., 1976. Seismicity and seismic hazard in Britain, Inst. of Geol. Sci. Seism. Bull. no. 4, HMSO, London.

Ringdal, F., 1982. Seismicity of the North Sea area, Present Volume: חꞋ-Ꞌꞎ

Ritsema, A.R., 1966. Note on the seismicity of the Netherlands, Proc. Kon. Ned. Akad. Wetens., Ser. B, 69, no. 2: 235-239.

Ritsema, A.R., 1980. On the Assessment of seismic risk in the North Sea area, Verslagen V-366, Kon. Ned. Meteor. Inst. De Bilt, pp. 19.

Ritsema, A.R., 1981. The seismicity of the North Sea Basin, Proc. Int. Symp. Analysis Seismicity, Seismic Hazard, Geoph. Inst., Czechoslovak Acad. Sci. Praha, Editor V. Schenk: 155-162.

Van Gils, J.M., Y. Zaczek, 1978. La seismicité de la Belgique et son application en génie parasismique, Ann. Trav. Publ. Belgiques, no. 6: pp. 39.

Van Rummelen, F.H. 1945. Overzicht van de tussen 600 en 1940 in Zuid Limburg en omgeving waargenomen aardbevingen, en van aardbevingen welke mogelijk hier haren invloed kunnen hebben doen gelden, Meded. Jaarverslag Geol. Bureau Heerlen 1942-'43, no. 15: pp. 130.

SEISMICITY AND NEOTECTONIC STRUCTURAL ACTIVITY OF THE RHINE GRABEN SYSTEM IN CENTRAL EUROPE

L. Ahorner

Abteilung für Erdbebengeologie, Geologisches Institut,
Universität zu Köln
Federal Republic of Germany

Abstract. A brief overview of the seismicity and neotectonic structural activity of the Rhine graben system is given, together with some new results of modern seismological research work which are of interest for a realistic assessment of possible seismic hazards in the Rhine area and its vicinity.

1. Neotectonic Block Movements

The Rhine graben system forms a tripartite (triple junction) rift structure in Central Europe with the three branches: Upper Rhine graben (URG), Lower Rhine graben (LRG) and Hessian graben zone (HGZ) (see Figure 1).

Only the Upper Rhine graben (between Basel and Frankfurt) and the Lower Rhine graben (from Bonn to the Netherlands) show a clear neotectonic (Quaternary and Recent) structural activity and a remarkable seismicity, whereas the Hessian graben zone (between Frankfurt and Kassel) seems to be rather inactive in post-Tertiary times.

The total amount of Quaternary vertical block movements along the border faults of the Upper Rhine graben exceeds 500 m. Geodetically observed present-day slip rates are in the order of 0,5 mm/year. In the Lower Rhine graben Quaternary deposits are dislocated along still active normal faults (e.g. Erft fault, Peel boundary fault) up to 175 m especially in the western part of the graben zone. Recent slip rates of about 1 mm/year have been proven by high precisions levellings. The amount of lateral crustal tension by which both flanks of the Lower Rhine graben have been drifted

101

A. R. Ritsema and A. Gürpınar (eds.), Seismicity and Seismic Risk in the Offshore North Sea Area, 101–111.

away from each other since early Quaternary times is approximately
100-200 m (Ahorner 1962).

The neotectonic structural activity of the Rhine graben system
is a typical example for tensional intra-plate tectonics. We may
understand it as secondary adjestive movements of smaller crustal
units (West Rhenish-Gallic block, North Rhenish-Lower Saxon block,
Southwest German block) within the Eurasian plate near the southern
boundary of this plate against the Mediterranean microplate mosaic
and the shifting African plate in the South.

2. Seismicity and Seismotectonics

The Upper Rhine graben and the Lower Rhine graben form together
the Rhenish seismoactive zone (see Figure 1). This zone, following
the course of the Rhine river from Basel in the South to the Nether=
lands in the North, is the most conspicuous seismological feature
in Central Europe north of the Alps. In the region west of Köln the
Rhenish seismoactive zone splits up and an important lateral branch
- the Brabant or Belgian seismoactive zone - strikes from here
westward across Belgium to the Channel coast near Ostend.

Earthquakes occurring along both zones have focal depths between
several km and 25 km as a maximum. The maximum observed magnitude
is about $M_L=6,5$ in the southernmost part of the Upper Rhine graben
(Basel earthquake of 1356 with intensity $I_O=IX-X$ MSK-scale) and
$M_L=6$ in the Lower Rhine graben (Düren earthquake of 1756 with
intensity $I_O=VIII$ MSK-scale) and in the Belgium seismoactive zone.

For the Lower Rhine graben and the adjoining parts of the Rhenish
massiv a new magnitude-frequency diagram has been derived by com=
bining historical earthquake data with the results of microearth=
quake investigations in the period 1977-1979 (see Figure 2). In
the magnitude range $1 < M_L < 5,5$ the relationship between local
magnitude M_L and frequency of occurrence log N/year is quite linear
with a b-value of 0,74. A similar b-value has been found by Bonjer
(1980).

From near-field digital recordings of local earthquakes in the
Rhine area in the period 1977-1982 the source characteristics of
events with local magnitudes M_L between 0,9 and 4,7 have been
estimated using S-wave spectra and the Brune source model. Stress
drops variing between a few tenths of a bar and 70 bar have been
found. The relationship between seismic moment M_O (dyne-cm) and
local magnitude M_L derived from our results for the Rhine area is
well described by the formula (see Figure 3):

$$\log M_O = 17,4 + 1,1 \ M_L$$

The standard deviation for log M_O is 0,21. A similar relationship has been found by Johnson & McEvilly (1974) for Central California earthquakes near the San Andreas fault and by Durst (1981) for the southern part of the Upper Rhine graben.

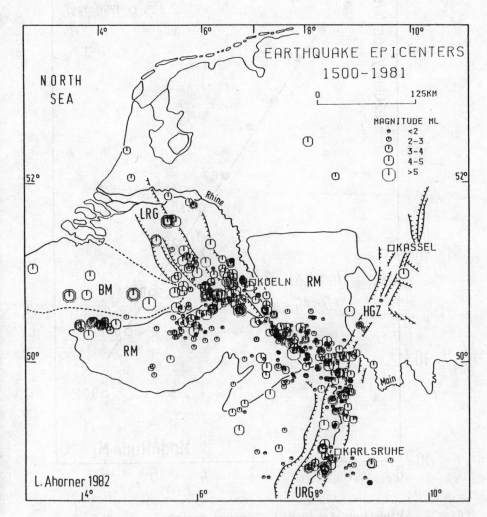

Fig. 1   Epicenter map of the Rhine area and the fault pattern
         of the Rhine graben system. More than 1000 earthquake
         epicenters are plotted with local magnitudes between
         1< M_L< 6. URG = Upper Rhine graben, LRG = Lower Rhine
         graben, HGZ = Hessian graben zone, RM = Rhenish massiv,
         BM = Brabant massiv

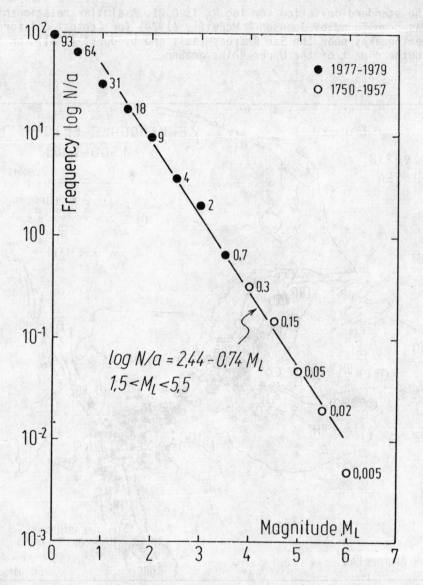

Fig. 2   Magnitude-frequency diagramm for the Lower Rhine graben and the adjoining parts of the Rhenish massiv. The area covered is about 40 000 km2. Open circles are data points from historical earthquakes, black circles from microearth= quake investigations in the years 1977-1979.

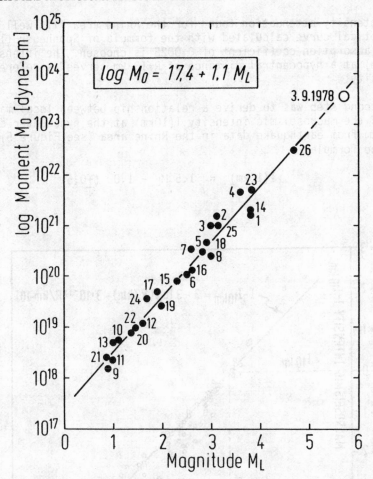

$$log\ M_0 = 17,4 + 1,1\ M_L$$

Fig. 3 Relation between seismic moment log M_O and local magnitude
        M_L derived from local earthquakes in the Rhine area. The
        data point of the Albstadt earthquake 1978 (Swabian Alb,
        magnitude M_L=5,7, intensity I_O=VII-VIII) is near the calcu=
        lated regression line thus indicating that the relation
        found fits Central European events in a wide magnitude range.

Important with respect to a realistic seismic hazard analysis is
a specific magnitude and hypocentral distance dependency of macro=
seismic intensity which has been derived from local events in the
Rhine area. In a first step the attenuation of macroseismic intensity
(MSK-scale) with hypocentral distance has been determined from an
analysis of 19 earthquakes observed in the Rhine area between 1864
and 1965 (see Figure 4 and Ahorner & Rosenhauer 1978).

The intensity attenuation found for the Rhine area fits well the theoretical curve calculated with the formula of Sponheuer (1960) if an absorption coefficient of 0,0025 is choosen. The intensity I(10km) at a hypocentral distance of R=10 km serves as reference intensity.

The second step was to derive a relationship between local magnitude M_L and the macroseismic intensity I(10km) at the hypocenter distance R=10 km from earthquake data in the Rhine area (see Figure 5). We get the formula

$$I(10km) = 1,5 M_L - 1,0 \quad (\pm 0,6)$$

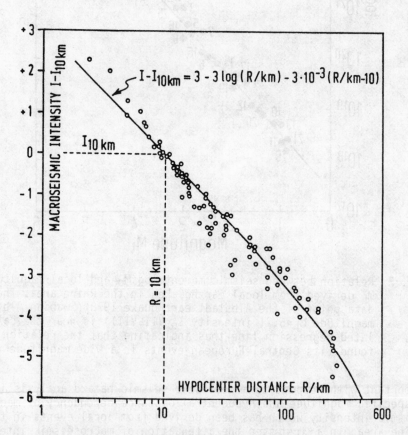

Fig. 4   Attenuation of macroseismic intensity I (MSK-scale) with hypocentral distance R. Points give mean isoseismal radii for 19 earthquakes observed in the Rhine area from 1846-1965 (Ahorner & Rosenhauer 1978)

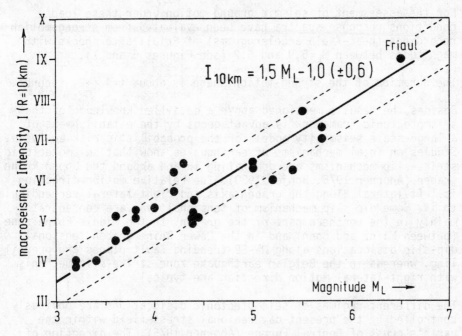

Fig. 5 Relationship between local magnitude M_L and macroseismic intensity $I(10km)$ observed at 10 km hypocentral distance. Data points are mostly from local earthquakes in Central Europe with well determined instrumental magnitude M_L and accurate macroseismic observations.

Combination of the results of step one and two yields the desired intensity attenuation law valid for the Rhine graben system and adjacent regions:

$$I(M_L,R) = 1,5 M_L + 2 - 3 \log R - 1,3 \propto (R - 10)$$

I is the macroseismic intensity (MSK-scale) at the site with the hypocentral distance R in km and $\propto$ the absorption coefficient in km^{-1} (mean value for the Rhine area 0,0025 km^{-1}).

With this attenuation law a realistic seismic hazard analysis for sites in the Lower Rhine area has been performed (see the contri= bution of W.Rosenhauer).

From the historical earthquake activity and from probabilistic risk studies on the basis of a large-scale seismicity model for the Rhine graben area it becomes clear that earthquakes with $M_L \sim 6$ and R=15-25 km are the most probable maximum event for many sites.

For the assessment of seismic ground motion under these load
conditions response spectra have been evaluated from strong-motion
recordings (near-field accelerograms) of Friuli aftershocks with
magnitudes between M_L=5,9 and 6,2 (see Figures 6 and 7).

The duration of the strong-motion phase is about t=4 ($\pm$2) seconds.

Besides the results mentioned above a detailed knowledge of the
seismotectonic framework is advantageous by the establishment of
a large-scale seismicity model for the probabilistic risk analysis.
Studies on focal mechanisms of earthquakes show that seismotectonic
strike-slip mechanisms are dominating in and around the Upper Rhine
graben (Ahorner 1975, Bonjer 1980). The relative motion direction
is left-lateral along the graben axis and right-lateral perpendicular
to it. Some dip-slip mechanism of tensional type are concentrated
mainly in the northern part of the graben. In the Middle Rhine zone
(between Mainz and Bonn) and in the Lower Rhine graben tensional
dip-slip dislocations along NW-SE trending fault planes are prevail=
ling, whereas in the Belgian earthquake zone strike-slip mechanisms
with right-lateral motion direction are typical.

The different regimes of seismotectonic dislocations are obviously
controlled by the present-day regional stress-field within the
earth's crust of Central Europe (Ahorner 1975). The direction of
maximum compressive stress as deduced from in-situ stress measurement
(Illies et al.1981) and earthquake focal mechanisms is NW-SE, the
direction of minimum compressive stress or tensile stress is SW-NE.
Under the influence of this tectonic stress field the NNE trending
Upper Rhine graben is affected mainly by horizontal shear forces
because the graben axis lies in the direction of maximum shear
stress (see Figure 8). Similar stress conditions are given for the
WNW trending Belgian earthquake zone. The NW trending Middle Rhine
zone and the Lower Rhine graben, on the contrary, is arranged
transverse to the direction of minimum horizontal stress. Therefore
mainly tensional forces perpendicular to the graben axis are acting
here.

From geological evidence it becomes clear that the tectonic stress
field has changed its direction during the long time of the tectonic
evolution of the Rhine graben system (see Figure 8). In Early Tertiar
time when the taphrogenesis of the Upper Rhine graben has started
the compressive stress direction was NNE. Later this paleo-stress
field has rotated anticlockwise by about 60° until the present
position was reached. The scheme of Fig. 8 demonstrates in which
manner the tectonic evolution of the Rhine graben system in its
entirety might have been influenced by the progressively rotating
stress field.

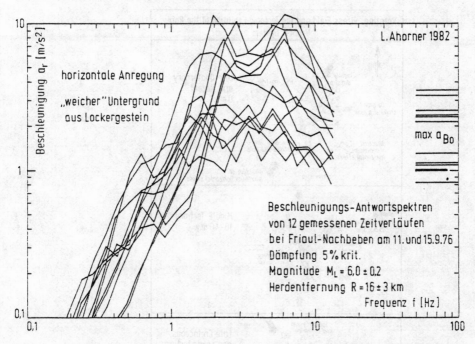

Fig. 6 Response spectra of Friuli aftershocks with M_L = 5,9-6,2

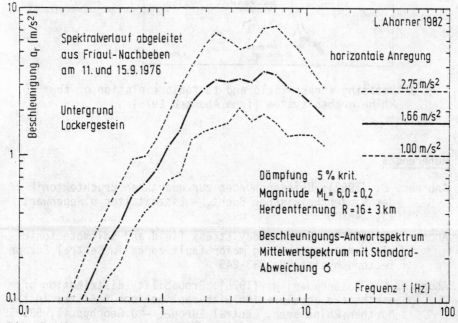

Fig. 7 Mean spectrum with standard deviation deduced from Fig.6

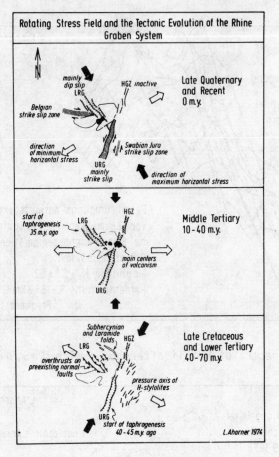

Fig. 8   Rotating stress field and tectonic evolution of the
         Rhine graben system (from Ahorner 1975)

References

Ahorner, L. (1962): Untersuchungen zur quartären Bruchtektonik
        der Niederrheinischen Bucht. - Eiszeitalter u.Gegenwart
        13, 24-105

Ahorner, L. (1975): Present-day stress field and seismotectonic
        block movements along major fault zones in Central Europe.
        Tectonophysics 29, 233-249

Ahorner, L., Rosenhauer, W. (1975): Probability distribution of
        earthquake acceleration with applications to sites in the
        Northern Rhine area, Central Europe. - J.Geophys.41, 581-594

Ahorner, L., Rosenhauer, W. (1978): Seismic risk evaluation for
        the Upper Rhine graben and its vicinity. - J.Geophys.44,
        481-497

Bonjer, K.-P. (1980): The seismicity of the Upper Rhinegraben -
        A continental rift system. - Proceedings of Int.Conf.on
        continental earthquakes, Ohrid 1979 (in press)

Durst, H. (1981): Digitale Erfassung und Analyse der physikalischen
        Prozesse in den Erdbebenherden aus dem Bereich des Ober=
        rheingrabens. - Diplomarbeit Universität Karlsruhe

Illies, J.H., Baumann, H., Hoffers, B. (1981): Stress pattern and
        strain release in the Alpine foreland. - Tectonophysics
        71, 157-172

Johnson,L.R., McEvilly, T.V. (1974): Near-field observations and
        source parameters of Central California earthquakes. -
        Bull.Seism.Soc.Am. 64, 1855-1886

Sponheuer, W. (1960): Methoden zur Herdtiefenbestimmung in der
        Makroseismik. - Freib.Forschg.Hefte C 88, 23-52

Van Gils, J.-M., Zaczek, Y. (1978): La séismicité de la Belgique
        et son application en génie parassismique. - Ann.des
        Travaux Publics de Belgique, No.6, 1-38

Rosenhauer, W. (1982): Methodological aspects encountered in the
        Lower Rhine area seismic hazard analysis. - (this volume)
        385-396.

REMOVAL OF GHOSTS

Jørgen Hjelme

Geodætisk Institut, Copenhagen, Denmark

A panel contribution:

After the excellent presentation by Dr. Ringdal very litle can be added to the seismicity in the North Sea area. But a few words of warning can be said in general.

Much care must be excercised when using existing catalogues when even small events can influence the total picture. As an example we shall use one of the best selections of epicentre determinations in the world, the ISC bulletin.

An earthquake happened in 1978 in the North Sea. In the ISC you will find two epicentres with very short time interval, but with most of the North Sea inbetween. The western point is determined by four readings from stations in Scotland together with P-arrivals at Moxa and Collm. A closer inspection revealed no station was reporting both earthquakes.

Table 1 is a list of readings from both events compared with the eastern hypocentre. The readings from the western proposal are marked with a star. The residuals O-C for P are related to the Jeffreys-Bullen tables. A few O-C for Sn according to a table by Båth have been added by hand. The readings from Umeå (UME) and Sodankylä (SOD) were treated as Sn already by the

113

A. R. Ritsema and A. Gürpınar (eds.), Seismicity and Seismic Risk in the Offshore North Sea Area, 113–114.
Copyright © 1983 by D. Reidel Publishing Company.

Table 1.

1978 apr 26

Estimated epicenter
56.55 N   8.20 E h= 10 km
H: 12 32 30.40
elo was not found in catalog
✳ qhb was not found in catalog

		Dist.	Az.S	Az.E	J·B P o-c	Båth S_n o-c	
kon	12 33 22.7	3.20	-166	13	+0.9		
ber	12 33 34.0	4.13	157	-20	-1.1		
hfs	12 33 43.1	4.61	-139	37	+1.2		
upp	12 34 03.4	5.98	-120	52	+2.1		
egl	12 34 02.1	6.14	79	-92	-1.4		✳
edu	12 34 03.3	6.20	85	-85	-1.1		
eka	12 34 07.0	6.50	75	-96	-1.5		✳
eau	12 34 07.6	6.54	79	-91	-1.6		✳
ebh	12 34 07.1	6.51	82	-88	-1.6		✳
esk	12 34 11.0	6.53	74	-96	+2.0		
dou	12 34 12.2	6.82	17	-160	-0.8		
mox	12 35 10.0	6.25	-18	160	+64.9	-4.8	✳
cll	12 35 14.0	5.97	-26	150	+73.0	+2.3	✳
ume	12 36 28.9	9.42	-135	34	+99.7	-3.4	
nur	12 34 50.0	9.47	-108	58	+0.2		
kjf	12 35 24.0	12.28	-120	43	-4.1		
sod	12 38 07.0	13.81	-133	31	+138.6	-11.1	

ISC. Inspection of the Collm station bulletin shows the
station analyst has interpreted the arrival as Sn. If
we also accept the arrival at Moxa as Sn, all readings
are explained by the eastern epicentre. The western one
must be a ghost earthquake.

We have thus lost one of the few earthquakes in
the North Sea.

COMMENTAIRES SUR UNE CARTE SEISMIQUE DE LA MER DU NORD ET DE LA MER DE NORVEGE*

E. Peterschmitt

Centre Séismologique Europeo-Méditerranéen, Strasbourg

Abstract -

The conditions for establishing of seismicity map are discussed : reliability of epicentral location from inadequate geographic distribution of data, magnitude definition.

Apart from the June 7, 1931 earthquake (the most important computed event), the seismicity of the North Sea and Norwegian Sea is located within a stripe paralleling the coast of Norway.

La base de la carte séismique présentée est constituée par un extrait de la carte générale des déterminations épicentrales réalisées par le C.S.E.M., pour la période 1976 à 1981. Cette base a été complétée pour la région de la mer du Nord et de la mer de Norvège par des informations provenant de diverses sources.

Déterminations épicentrales

1) Déterminations du C.S.E.M. (Tableau I)

Les objectifs fixés au C.S.E.M. lui imposent de calculer et de diffuser les résultats dans un délai de 2 à 3 mois après les évènements. Il en résulte que les données utilisées ne sont pas complètes et que,pour des séismes faibles,des différences non négligeables peuvent exister entre les résultats du C.S.E.M. et ceux, en principe plus complets, de l'International Seismological Centre. Sauf divergences trop importantes, nous avons conservé les résultats du C.S.E.M.

*Presented by Dr. Hoang Trong Pho.

A. R. Ritsema and A. Gürpinar (eds.), Seismicity and Seismic Risk in the Offshore North Sea Area, 115–120.
Copyright © 1983 by D. Reidel Publishing Company.

Remarques :

a) 24.2.1978 : il n'est pas possible de faire un choix entre les épicentres de l'I.S.C. - 62.23 N , 1.81 E - et celui du C.S.E.M. - 61.37 N , 2.61 E - ; les deux résultats sont statistiquement équivalents suivant la préférence donnée aux temps d'arrivées de telle ou telle station.

b) 28.2.1978 : De nouveaux calculs conduisent à un épicentre voisin de celui de l'I.S.C.

c) Ces deux exemples montrent les difficultés qui peuvent se présenter. Le manque de fiabilité de certains résultats est la conséquence d'une mauvaise distribution des stations autour des foyers. Il faut tenir compte de ce fait pour les analyses ultérieures.

Les déterminations des profondeurs focales doivent être considérées, sauf exception, comme illusoires.

2) Autres déterminations (Tableau II)

Les remarques ci-dessus nous ont guidés pour la rédaction du Tableau II - période 1902 à 1975 -. Ce tableau n'est pas complet : seuls ont été retenus les évènements assez importants pour qu'une fiabilité minimum des résultats soit assurée.

Seuls ont été reportés sur la carte les tremblements de terre de magnitude supérieure à 4,8 pour les déterminations de l'I.S.C. - période 1966-1975 -, et 5 respectivement classes d et e de Gutenberg - pour la période précédente (1902-1965). Dans certains cas V. Karnik donne 2 déterminations. Elles ont été reportées toutes les deux et reliées par des tirets.

Problème des magnitudes

Pour les données du C.S.E.M. et celles avant 1965, la préférence a été donnée aux magnitudes d'ondes superficielles - ML ou MS - par rapport aux valeurs de MB. Le nombre de stations ayant fourni des observations de temps d'arrivée a permis de corriger certaines anomalies.

La carte séismique

L'évènement dominant la séismicité de la mer du Nord est naturellement le tremblement de terre du 7 juin 1931. La localisation adoptée par Karnik - 54.0 N , 1.4 E - , d'ailleurs assez proche de celle de l'I.S.S. - 53.8 N , 1.2 E - , est sans doute voisine de la réalité. La valeur de magnitude 6.0 indiquée par le même auteur est probablement une valeur minimum.

Tenant compte des difficultés de localisation, une zone pa-
rallèle à la côte norvégienne et distante d'elle de 200 km envi-
ron paraît être l'élément de risque séismique le mieux défini.

Enfin la région voisine de 66° N , 0° s'est manifestée à
2 reprises.

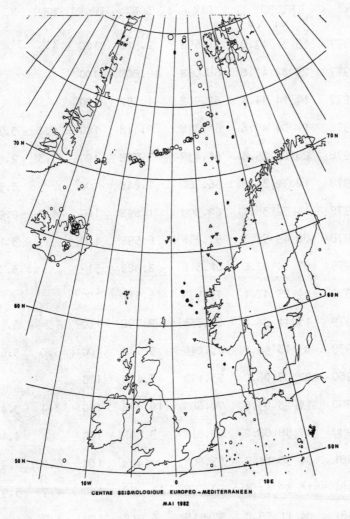

CENTRE SEISMOLOGIQUE EUROPEO - MEDITERRANEEN

MAI 1982

CRETE MEDIANE DE L'ATLANTIQUE : CSEM 1976 - 1981
MER DU NORD ET MER DE NORVEGE

| ▽ : Karnik | 1901 - 1955 | M ≳ 5 | □ : I.S.C. | 1966 - 1975 | M ≳ 4.8 |
| △ : Rothé | 1955 - 1965 | " | ● : C.S.E.M. | 1976 - 1981 | |

TABLEAU I

CENTRE SEISMOLOGIQUE EUROPEO-MEDITERRANEEN

Tremblements de terre localisés en Mer du Nord et Mer
de Norvège en dehors de la Crète Médiane de l'Atlantique

1976 - 1981

DATE	HEURE h mn s	LAT	LONG	PROF	MB	MS	NST
6. 4.1977	19 31 47.0	61.69N	2.31E	10C	4.5		48
30. 4.1977	23 32 43.8	68.08N	10.80E	10G			27
9.11.1977	14 14 44.4	62.90N	1.95E	10G			13
24. 2.1978	11 04 45.6	61.37N	2.61E	10G	4.2	3.7	15
28. 2.1978	20 23 25.4	61.43N	3.79E	10G	3.3	2.9	14
26. 4.1978	12 32 34.6	56.78N	7.84E	10G		3.3	12
19. 9.1978	14 52 36.4	62.30N	1.69E	10G	4.7	2.9	43
2.10.1978	08 49 54.3	57.75N	6.59E	10G		3.1	9
12. 5.1979	07 14 02.6	61.31N	2.98E	51		3.3	24
28. 6.1979	04 25 47.1	70.53N	16.51E	10G			15
28. 7.1979	15 41 34.8	61.35N	2.79E	10G		3.1	7
14.12.1979	03 13 40.8	65.04N	5.14E	10G	5.0	3.6	10
20. 1.1980	22 07 06.0	73.17N	13.46E	10G			46
29. 3.1980	14 13 03.9	70.68N	16.84E	10G	3.4	3.4	11
8. 6.1980	07 58 08.2	60.81N	3.60E	10G		4.1	23
22. 6.1981	04 38 14.5	65.86N	0.40E	10G	4.0		32
3. 9.1981	18 39 44.3	69.39N	14.27E	10C	4.7	3.9	76
6. 9.1981	04 11 58.0	57.11N	6.81E	10G	4.7	4.1	35
7. 9.1981	07 52 11.5	72.44N	14.23E	10G	4.0		18

TABLEAU II

Principaux tremblements de terre
originaires de la Mer du Nord et de la Mer de Norvège

DATE	HEURE h mn s	LAT	LONG	PROF	MS	SOURCE
9. 2.1902	02 59	59.5N	4.0E		(4.6)	K
6. 2.1905	17 09.7	62.0N	5.0E		(4.3)	K
* 14. 1.1907	13 03	65.5N	11.0E		5.1	K
* 27. 1.1907	04 58	65.5N	11.0E		(5.2)	K
* 30. 6.1908	04 53	67.3N	14.3E		5.0	K
24. 8.1911	21 48.4	60.0N	5.2E		(4.4)	K
*(19. 7. 1913	15 50.2	64.0N	8.0E		5.0)	K
(19. 7.1913	15 49.6	64.7N	6.2E		5 1/2)	
5. 5.1923	03 08.8	62.4N	6.0E		4.9	K
25. 7.1924	19 36 22	72.5N	16.0E		4.6	K
*(24. 1.1927	05 18 22	59 N	3 E		5.1	K
(	05 18 11	58.5N	6.0E			
15. 6.1927	06 16 21	61.7N	4.4E		(4.2)	K
23. 5.1929	18 36 18	57.5N	7.4E		4.5	K
29. 5.1929	23 31 05	57.7N	7.3E		4.2	K
* 10.6.1929	23 03 14	71 N	10 E		6.0	K
* 7. 6.1931	00 25 21	54.0N	1.4E		6.0	K
*(17. 7.1935	00 04 19	65.9N	7.2E		5.0	K
(	00 04 13	66.0N	8 E			
(27.11.1937	20 10 31	73.2N	10.2E		5.1	K
(	20 10 32	71 N	9 E		5	
11.03.1938	16 08 20	61.9N	4.2E		4.8	K
(11.05.1946	16 25 30	66 N	0.5W)		5.0)	K
	16 25 17	67.0N	5.0W)?		4 3/4)	
*(11.05.1946	18 39 27	66 N	0.5W		5.1)	K
(	18 39 25	65 3/4N	1 W		5 3/4)	
*(23.08.1948	11 50 39	72 N	4 E)?		5.1)	K
(	11 50 30	71.7N	10.0E)		5)	
* 4.11.1953	01 35 57	71.2N	9 E		5.0	K
* 3. 6.1955	11 39 35	62.1/4N	5 E		5.0	K
* 23. 1.1958	13 35 04	65.2N	6.8E	n	5.5	R
6. 8.1958	17 16 07	59.6N	5.8E	n	e	R
* 29. 1.1959	23 24 30	70.9N	7.3E	n	6	R
31. 5.1960	03 54 46	73.6N	14.6E	n	e	R
18.10.1962	10 35 41	60.9N	11.9E	n	e	R
* 5.11.1962	11 46 11	66.6N	7.1E	n	d	R
* 15. 8.1963	23 57 03	69.8N	8.9E	n	5.5	R
14. 7.1964	05 33 55	57.0N	7.3E	36	e	R
24.10.1965	12 16 59	57.0N	7.4E	n	e	R

DATE	HEURE h mn s	LAT	LONG	PROF	MB	NST	SOURCE
30. 6.1966	21 38 18	62.5N	4.9E	0	–	11	
11. 7.1966	21 37 49	67.29N	10.20E	41	4.1	15	
4. 9.1966	08 40 16	62.8N	6.2E	33	4.0	12	
*21. 8.1967	13 41 49	57.06N	4.92E	33	5.2	47	ISC
22. 6.1967	22 05 51	67.78N	10.6 E			13	ISC
25. 3.1968	11 28 20	70.81N	7.4 E	0	4.2	17	ISC
20. 7.1968	16 52 38	61.7 N	5.4 E	0		10	ISC
7. 9.1968	13 35 49	62.3 N	5.2 E	33		14	ISC
7.10.1968	06 55 14	61.4 N	4.0 E	0		13	ISC
21. 1.1969	21 09 56	73.78N	14.2 E	10	4.6	28	ISC
5. 4.1969	19 09 48	57.13N	7.04E	33	4.2	24	ISC
*29. 9. 1969	10 27 50	65.10N	6.5 E	6	4.8	65	ISC
28.11.1970	23 29 02	70.6 N	8.5 E	0	–	9	ISC
22.12.1970	07 59 49.6	71.42N	13.4E	0	–	12	ISC
23.03.1971	20 05 04.5	59.75N	2.3 E	0	–	17	ISC
20.08.1971	19 06 27.2	61.65N	4.7 E	35	4.2	36	ISC
08.11.1971	23 24 44.4	62.95N	5.1 E	33		11	ISC
06.12.1971	23 36 30.0	65.79N	7.7 E	45	4.3	15	ISC
21.04.1972	13 33 27	62.90N	2.5 E	45	(4.9)	15	ISC
12.12.1972	07 42 27.0	71.6 N	11.9 E	33	4.2	30	ISC
6. 3.1973	12 45 00	65.77N	0.10E	0	4.1	10	ISC
7. 8.1974	03 53 36	70.7 N	10.2 E	33		12	ISC
8. 8.1974	10 11 06	72.9 N	7.2 E	33		13	ISC
2.11.1974	22 34 22	73.32N	14.4 E	33	4.3	14	ISC
9.11.1974	07 14 24	69.64N	9.7 E	33		13	ISC
18.12.1974	20 12 16	67.85N	10.5 E	33	4.3	25	ISC
03.04.1975	06 39 17.3	59.33N	5.1 E	0		20	ISC
12.11.1975	00 06 16.0	57.01N	7.17E	95		36	ISC

Sources des déterminations

K : V. KARNIK, Seismicity of the European Area, Reibel, Holland, 1969.

R : J.P. ROTHÉ, La Séismicité du Globe 1953–1965, Unesco Paris 1964.

ISC : International Seismological Centre (Edinbourg, Newbury).

----

* : Epicentres reportés sur la carte

DISCUSSION-NOTES SECTION 2

Seismicity

Q. C.W.A. Browitt:
In my paper I showed a slide of epicentres of small North
Sea earthquakes detected by recently installed instruments
for the period 1980 to February 1982. Whilst these were
preliminary results, with variable error bars depending
on the availability of stations in both Britain and Norway,
a linear trend was observed extending into the mid North
Sea from the "knee" in the Norwegian coastline at $62^{\circ}N, 5^{\circ}E$.
This was possibly consistent with historical data presented
by Dr. Ringdal but the trend was orthogonal to one on the
EMSC map presented by Dr. Hoang. The EMSC data was said
to have latitude errors of about 1° with much smaller
errors in longitude. This situation of relatively well
determined azimuth with poor range control is common for
earthquakes on the edge of a regional network of observing
stations and it may give false lineations.

A. F. Ringdal:
It would be very difficult to use historical earthquake
data to identify a lineament of the type mentioned, because
of generally poor epicentral accuracy. Even for the most
recent period, 1951-1980, our data are not sufficiently
accurate in this respect. Thus an eventual confirmation
of Dr. Browitt's observation will have to await further
data from the relatively dense, land based seismograph
networks that are now in operation.

A. Hoang Trong Pho:
I have never said that the EMSC results show errors in
latitude of about 1° and smaller errors in longitude. I
have just mentioned that the divergence between ISC and
EMSC results is due to different combinations of data
used in epicentral determinations. And I gave the example
of the February 24, 1978 earthquake. Its coordinates are
$62^{\circ}23$ N, $1^{\circ}81$ E by ISC and $61^{\circ}37$ N, $2^{\circ}61$ E by EMSC. This
important difference in spatial parameters is typical of
a lack of reliability of some results when the gap in
azimuth and in distance is large.
I entirely agree with Dr. Browitt's remarks about false
lineations. Sometimes, these ones are simply the image
of a systematic orientation of the confidence ellipse.

121

A. R. Ritsema and A. Gürpinar (eds.), Seismicity and Seismic Risk in the Offshore North Sea Area, 121–123.
Copyright © 1983 by D. Reidel Publishing Company.

Q. A. Gürpinar:

We have observed in Turkey from earthquakes with variable I_o values, that intensity attenuation is significantly dependent on I_o. For $I_o \leqslant$ VII the near field attenuation ($R \sim 30$-50 km) is more pronounced than for earthquakes with $I_o >$ VIII. Therefore, the inference of I_o values for offshore epicenters from onshore observation may underestimate the actual values if the attenuation relationship is based on low I_o earthquakes.

A. F. Ringdal:

I would comment in general that near-field strong motion attenuation is not well understood at present. Your observations from Turkey are certainly interesting in this regard. Even though we have no similar data for the North Sea area, your conclusion might well be applicable, and would imply that a certain degree of conservatism should be introduced in assessing near-field effects from earthquakes in this area.

Q. A. Gürpinar:

The intensity values observed in the coastal areas of the North Sea are based upon high frequency (~ 10 hz) structures such as stone masonry. Therefore, their use for offshore structures (~ 0.3-0.4 hz) should include considerations of frequency content. In particular, intensity observation on high frequency strcutures would underestimate the intensity values on long period structures at long epicentral distances.

A. F. Ringdal:

Here, you touch upon a central question for offshore structures. Clearly, the most important frequencies with respect to earthquake risk for these structures are quite low, and the scaling of the design spectrum at low frequencies is of critical importance. I agree completely that large earthquakes, even at long epicentral distances, may be of importance here, and this is an area where I think further research needs to be done.

Q. Hoang Trong Pho:

Do you think there may exist some induced seismicity in the North Sea? In that case, the seismic risk near the offshore sites is created by the hydrocarbon exploration itself!

A. F. Ringdal:

I know of no definite evidence of induced seismicity from hydrocarbon extraction either in the North Sea or other

similar areas. I would not, however, completely rule out
the possibility that such induced seismicity might occur.

Q. C.W.A. Browitt:

I question the remark made by Dr. Hjelme regarding the
value of assessing the cost of seismological research in
relation to the value to industry of the risk assessments.
To do justice to such a comparison it would be necessary to
quantify the value of the research to
a) general human knowledge;
b) the understanding of earth processes which may have
   applied benefits elsewhere, and
c) the benefit of present research to future exploitation
   of North Sea resources beyond the oil and gas extraction
   phase.

A. J. Hjelme:

I can fully agree with the remarks by Dr. Browitt. Scientific work should be justified alone for its value for
Science. My thoughts went the other way round. It could
be profitable for the oil industry to support enhancement
of our knowledge in the North Sea because even large
amounts for scientists are still only fractions of the
cost involved in the offshore operations.

Section 3

Tides, Ocean Waves and Sea Level Changes

TIDAL RESEARCH AND SEISMICITY

Paul MELCHIOR

Royal Observatory of Belgium
Avenue Circulaire 3, 1180 Bruxelles

ABSTRACT

Several aspects of the tidal deformations may be of interest in seismicity analyses : possible earthquake triggering, dilatancy monitoring, precursor observations. Oceanic tides even may be more important than solid earth tides provided their loading effects.

While instrumentation should more and more be set towards real time observations, it should be great if more young scientists could be involved in seismological research in Europe.

Tidal research and observation may be relevant to seismology in the following three aspects.

1 - It is a rather popular belief that tidal stresses in the crust can trigger earthquakes. As regards this many papers have been published in order to prove the existence of such a correlation. More, moonquakes having undoubtly revealed a tidal effect, the problem for the earth has been revived deeply.

2 - As well known, the dilatancy process affects the velocity of seismic waves and thus appears as one of the most important precursors of earthquakes. Obviously the tidal deformations should be affected in a similar way. Beaumont and Berger (1974) indeed have shown that when the ratio V_p/V_s, for waves travelling through a region affected by dilatancy, may vary by some 15 %, then the corresponding effect in tidal tilt and

A. R. Ritsema and A. Gürpınar (eds.), Seismicity and Seismic Risk in the Offshore North Sea Area, 127–130.

stress might be of the order of 60 %. Such observations will
have the great advantage as regards the seismic phenomena
that tidal deformations will be a permanent process allowing
continuous monitoring of dilatancy.

3 - As the tidal instrumentation (as gravimeters, tiltmeters and
extensometers) generally has a natural period in the low frequency domain, these instruments may be considered as good
detectors for some typical forerunners of earthquakes. However, up to now, no clear example can be given.

In order to make clear that the above mentioned applications raise some difficulties, we should remind the main features of the tidal phenomenon for what concerns their amplitudes
and frequencies.

A fully detailed treatment of the tidal waves is given
in Melchior (1978).

Concerning the amplitudes of the tidal phenomena, their
magnitudes are in the order of
   0,5 m. in radial displacement (vertical), or
   0,2 mgal, i.e. 2.10^{-7} g in gravity,
   0″02 in tilt
   2.10^{-8} in cubic dilatation or linear strain, and about
   10 mbars in stress.

As regards the frequencies, these are mainly divided in
three "families" each of them having a different impact on the
rotation of the Earth :
- the semi-diurnal (sectorial)waves with a period of ca 12 h and
  is responsible for the secular retardation of the Earth's rotation;
- the diurnal (tesseral) waves, having a period of ca 24 h and
  responsible for the precession nutation of the Earth's axis of
  rotation;
- the 14 d, 28 d, 6 months and on year period or zonal waves implying periodic variations in the Earth's speed of rotation.

These different families are latitude dependent for
their three usual components (Vertical, NS and EW) what increases
the difficulties for setting out a correlation in time between
the earthquake occurrence and the maximum amplitude of the tide.
This indeed depends upon the chosen component and at which latitude the seismic activity is considered.

For these reasons it is easier to establish a correlation in tropical areas by taking the vertical component or the
cubical dilatation as the tidal curve for these components is a
very regular one at low latitudes.

Considering restricted areas separately (Flinn & Engdahl) interesting correlations can be shown (Van Ruymbeke et al. 1981).

However, not only the "solid body tide" is responsible for triggering earthquakes eventually. The oceanic tides too, having exactly the same frequencies, may induce earthquakes by their well-known loading effects.

Generated by the same lunisolar potential, the oceanic tides obey to dynamic equations being those of Navier Stokes on a rotating body (the Coriolis force playing a major rôle), and dissipate energy by kinematic friction on the boundaries and by turbulence as well.

The interaction between the solid earth tides undergone by the ocean floor and the loading due to the masses of water must be introduced in the equations.

This leads the problem of the numerical integration of the equations to a more intricate one that several authors tried to solve since about 20 years. As a result of such an integration, Schwiderski, in 1979, elaborated "cotidal maps" for the ten main tidal waves.

Each map consists in a set of some 45.000 small polygons (1° x 1°) representing the world's oceanic surface. For the centre of each polygon height and phase are given for each tidal wave. So, with the aid of these maps, it is possible to determine the loading at any point of the Earth by the Green functions procedure (Farrel 1972). Such maps are of great importance in the reduction of satellite altimetry measurements. As the oceanic tides are rather important in the area of the North Sea their loading effects are significant as well.

Since 1973, Transworld Tidal Gravity Profiles are carried out by the International Centre for Earth Tides (I.C.E.T.) at Brussels. A first one runs from Istanbul to Tahiti and contains not less than 50 measuring sites, while another one links Caïro to Maputo over 15 stations.

At each site, measurements were made during 6 months with gravimeters, the best available ones presently.

After completion of both these profiles, an analysis of the results obtained have shown that :
1 - the Schwiderski maps are excellent;
2 - the effect of to the earth's flattening and rotation on the tidal deformations have been observed;
3 - in areas as the Alps, the Himalayas, Indonesia, the South

Pacific and the Corner of Africa, lateral heterogeneities in
the lithosphere and upper-mantle have been set out by the
tidal deformations.

Nowadays a new type of high sensitive gravimeter has
been designed at the University of California (Prothero and
Goodkind 1972). This instrument, now available, is based upon the
supraconductivity properties of some substances : so a ball is
maintained in levitation by a magnetic field kept constant inside
a Dewar filled up with liquid helium. Two of these new gravimeters presently have been installed in Europe : one at Frankfurt
and another one at Brussels.

In remote sensing operation they could be used in real
time for detecting any tidal and non-tidal anomalies in the crust
under the Western european area.

Recommendations.

Considering the relations possible between tidal effects
and seismicity, it should recommended :
1 - to involve more young scientists for seismological research
    in Western european countries;
2 - to conceive and design more and more sensitive instruments;
3 - to set greater value upon the instrumental calibration;
4 - to monitor tidal deformations in real time.

References

Beaumont, C. and Berger, J., 1974. Earthquake Prediction - Modi-
        fication of the earth tide tilts and strain by dilatan-
        cy. Geoph. Journal Roy. Astr. Soc. 39 : 111-121.

Farrell, W.E., 1972. Deformation of the Earth by surface loads.
        Rev. Geophys. Space Phys., 10 : 761-797.

Melchior, P., 1978. The Tides of the Planet Earth. Pergamon Press,
        Oxford : 609 pp.

Prothero, W.A. and Goodkind, J.M., 1972. Earth tide measurements
        with the superconducting gravimeter. Journ. Geoph. Res.
        77, pp. 926-936.

Schwiderski, E.W., 1980. On charting global ocean tides. Rev.
        Geophys. Space Phys., 18.

Van Ruymbeke, M., Ducarme, B., and De Becker, M., 1981. Parame-
        trization of the tidal triggering of earthquakes. Bull.
        Infor. Marées Terrestres 81 : 5521-5544.

CHANGES OF RELATIVE MEAN SEA-LEVEL AND OF MEAN TIDAL AMPLITUDE ALONG THE DUTCH COAST

J.G. de Ronde

Directie Waterhuishouding en Waterbeweging, Rijkswater-
staat, The Hague, The Netherlands

1. CHANGE IN RELATIVE MEAN SEA-LEVEL

1.1. With data derived from tidal gauges

In figure 1 to 7 some relative sea-level curves are given, with their linear variations and nodal tide, from gauge-records along the Dutch coast. The locations of these gauges are shown on the map in fig. 8. Figure 9 allows the comparison of the stations concerned. The linear variation (relative sea-level rise) varies between 16 and 22 cm per century and appears to be higher in the southern part of the Dutch coast than in the northern part (table I). To calculate the mean sea level rise the mean

Table I: relative sea-level rise in cm per century

	this publ.		Rossiter	
		data used		data used
Flushing	22	1900...1980	30	1890...1962
Zierikzee	17	1900...1980	17	1874...1962
Hook of Holland	19	1900...1980	25	1874...1962
IJmuiden	21	1900...1980	24	1884...1962
Den Helder	16	1933...1980	15	1874...1962
Harlingen	16	1933...1980	14	1874...1962
Delfzijl	16	1900...1980	17	1874...1962

A. R. Ritsema and A. Gürpınar (eds.), Seismicity and Seismic Risk in the Offshore North Sea Area, 131–141.
Copyright © 1983 by D. Reidel Publishing Company.

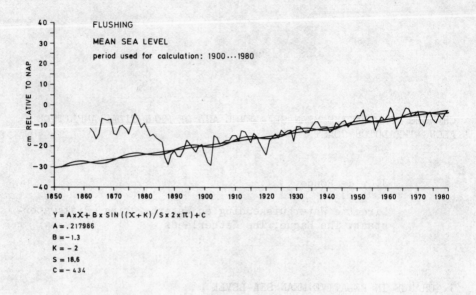

$Y = A \times X + B \times SIN ((X + K) / S \times 2 \times \pi) + C$
$A = .217986$
$B = -1.3$
$K = -2$
$S = 18.6$
$C = -434$

Figure 1

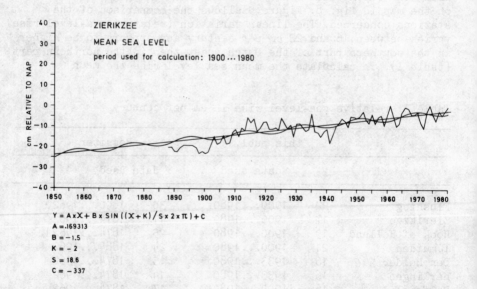

$Y = A \times X + B \times SIN ((X + K) / S \times 2 \times \pi) + C$
$A = .169313$
$B = -1.5$
$K = -2$
$S = 18.6$
$C = -337$

Figure 2

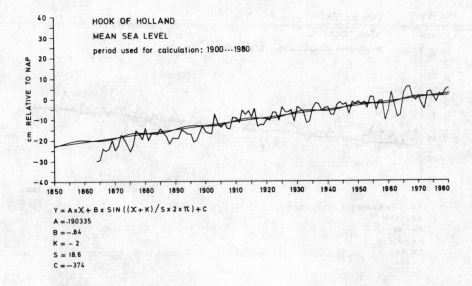

$$Y = A \times X + B \times SIN ((X+K)/S \times 2 \times \pi) + C$$
A = .190335
B = −.84
K = − 2
S = 18.6
C = −374

Figure 3

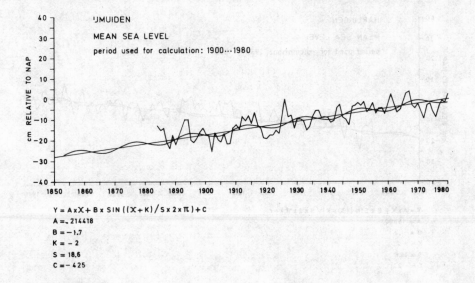

$$Y = A \times X + B \times SIN ((X+K)/S \times 2 \times \pi) + C$$
A = .214418
B = −1.7
K = − 2
S = 18.6
C = − 425

Figure 4

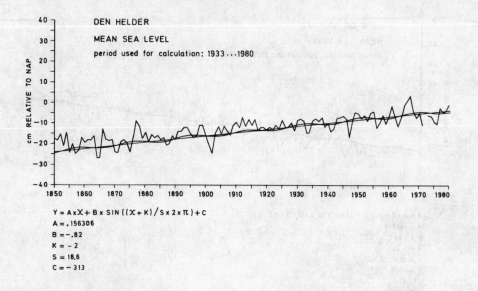

Figure 5

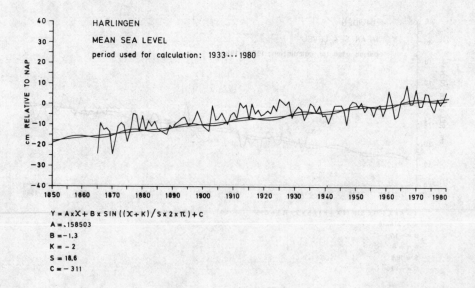

Figure 6

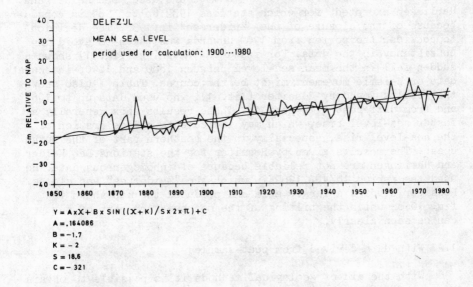

Figure 7

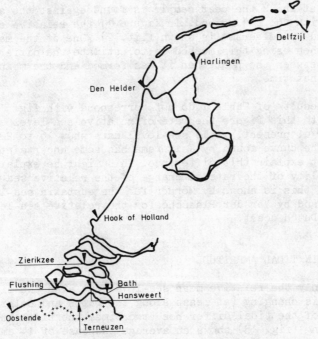

SITUATION OF GAUGES

Figure 8

sea-level per year from 1900 up to 1980 are used, Den Helder and
Harlingen excepted, for which stations 1933 up to 1980 is considered
because of the closure of the "Zuiderzee" in 1932. The decision
to consider the period from 1900 unwards was taken to avoid in-
consistency in the data. For instance: Flushing (fig. 1) shows a
sudden fall in the mean sea-level between 1880 and 1890; inaccurate
data or seismic movement might be the course. Table I also shows
the results found by Rossiter (lit. 1), who used data up to 1962
and from different time-periods. These results show generally a
faster sea-level rise, which may indicate to a deceleration of
the sea-level rise, especially in the southern part of the Dutch
coast. The results given by Rossiter for the stations Den Helder
and Harlingen are not reliable because of inhomogeneous data. He
used the period 1874...1962 in which the "Zuiderzee" was closed.
Rossiter studied a larger area and produced the map shown in
fig. 10 on which the tilting of the northwestern part of Europe
can be seen clearly.

1.2. With data derived from peat-samples

With the aid of geological methods it is possible to obtain
a curve for the relative mean sea-level from 8000 years BP (Before
Present) up to 3000 years BP. This curve can be found by plotting
the depth at which the peat sample is found against the age
measured with the C^{14} method. In figure 11 the relative sea-level
curve found by Jelgersma is shown (lit. 2). One of the main
problems when using this method is to estimate the differences
in level between the peat when it was formed and the mean sea-
level at that time.

The results of Table I do not correspond with fig. 11.
According to this figure the rate of relative sea-level change is
about zero at present, while sea-level data show 16 to 22 cm/
century. The curve found by Jelgersma has some uncertainties, but
this cannot explain this difference. This might be explained by
an instability of the rate of change of the relative sea-level.
In fig. 12 this is shown by Mörner for the eustatic sea-level in
NW-Europe and by Van der Plassche for the relative sea-level
along the Dutch coast.

2. CHANGE IN TIDAL AMPLITUDE

Not only the relative mean sea-level but also the tidal
amplitude is changing (at least along the Dutch and Belgian coast).
The graph of the tidal difference (two times the tidal amplitude)
for Flushing (fig. 13) shows an average increase of 14 cm per
century. Together with the relative sea-level rise this results
in a rise of the mean high water level with an average rate of
33 cm per century (fig. 14) and a rise of the mean low water-level

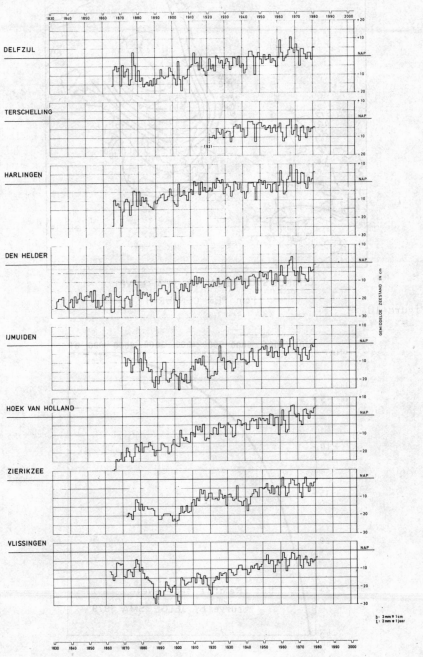

Figure 9

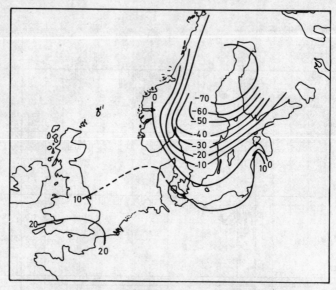

CHANGES OF RELATIVE SEA-LEVEL
(cm PER CENTURY)

Figure 10

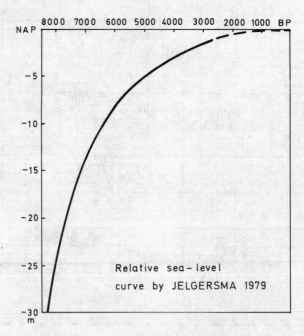

RELATIVE SEA-LEVEL CURVE

Figure 11

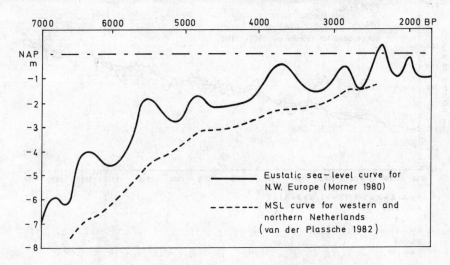

VARIATIONS OF THE RELATIVE SEA-LEVEL

Figure 12

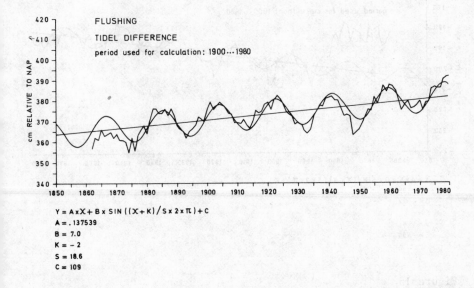

$$Y = A \times X + B \times SIN ((X+K)/S \times 2 \times \pi) + C$$

A = .137539
B = 7.0
K = -2
S = 18.6
C = 109

Figure 13

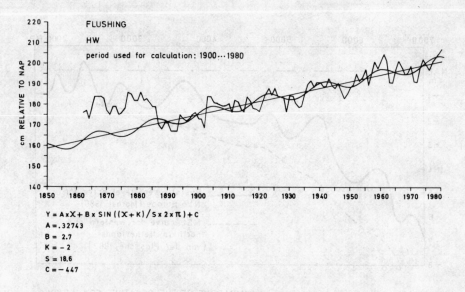

Figure 14

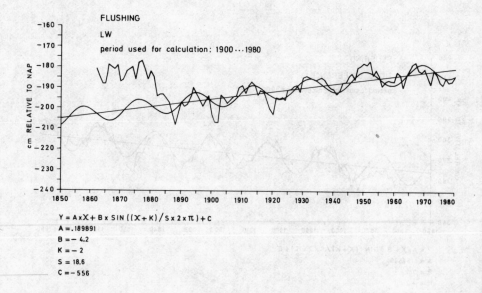

Figure 15

of 19 cm per century (fig. 15). In Table II the results are given
for some more stations.

Table II: average rate of rise of relative mean sea level (MSL),
mean high water level (HW), mean low water level (LW)
and mean tidal difference (TD) in cm per century.

Station	MSL	HW	LW	TD	period used
Oostende		17	5	12	1925-1980
Flushing	22	33	19	14	1900-1980
Terneuzen		40	18	22	1900-1980
Hansweert		40	10	30	1900-1980
Bath		44	16	28	1900-1980
Zierikzee	17	26	5	21	1900-1980
Hook of Holland	19	22	16	6	1900-1980
IJmuiden	21	24	18	6	1900-1980
Den Helder	16	15	7	8	1933-1980
Harlingen	16	27	10	17	1933-1980
Delfzijl	16	21	7	14	1900-1977

The results show an increase of the tidal difference along
the Dutch and Belgian coast at a rate of 6 to 21 cm per century.
This rate is the highest (up to 30 cm per century) along the
"Westerschelde", most likely because of dredging. Unfortunately
at present this increase of tidal amplitude along the Dutch and
Belgian coast cannot yet be explained.

Further study on the topics mentioned above is continuing.

REFERENCES

Jelgersma, S, 1979. Sea-level changes in the North Sea basin.
In: E. Oele, R.T.E. Schüttenhelm and A.J. Wiggers (eds.):
The quarternary history of the North Sea - Acta Univ. Ups.
Symp. Univ. Ups. Annum Quingentesium Celebrantis 2, 233-248.
(lit. 2)

Rossiter, J.R., 1967. An analysis of annual sea level variations
in European waters, Geophys. J.R. Astr. Soc., 12, 259-299.
(lit. 1)

EXTREME VALUES OF WIND SPEED AND WAVE HEIGHT IN THE NORTH SEA
FOR DEFINING DESIGN CRITERIA OF OFFSHORE INSTALLATIONS

by E. Bouws (KNMI, De Bilt)

J.J.E. Pöttgens (Staatstoezicht op de Mijnen, Heerlen)

ABSTRACT

Some climatological data are presented in terms of an average
recurrence period of 50 years, cited from a report of Working
Group I - Environment of the Conference on Safety and Pollution
Safeguards in the Development of North West European Offshore
Mineral Resources. The hourly mean wind speed in the northern
part of the North Sea may reach 38-40 m/s, which is about 20%
greater than the value found in the southern North Sea. On the
other hand, wave heights are likely to exceed the 30 m level in
the northern North Sea, which is about twice the value found in
the south. This is due to the combined effect of differences in
wind climate, water depths and fetch lengths.

1. INTRODUCTION

In the North Sea, in many cases wind and waves form the most
serious risk factors. This contribution hopefully serves to get
some feeling for this kind of risks compared to risks of seismic origin.
Traditionally, KNMI is active in the field of marine meteorology.
In the past, its nautical department collected many meteorological and oceanographic data from ships' logbooks. At present, the
division of oceanographic research is continuing this work, together with physical oceanographic research more or less related
to it.
One of the standard subjects of oceanographic research at KNMI

143

A. R. Ritsema and A. Gürpınar (eds.), Seismicity and Seismic Risk in the Offshore North Sea Area, 143–148.
Copyright © 1983 by D. Reidel Publishing Company.

is ocean waves. This comprises both wave modelling and climatological research, sometimes interacting with each other.
Encouraged by the Mine Inspection there has been a joint climate
study by KNMI and the offshore industry aimed at the description
of the wind and wave climate of the Dutch sector of the North Sea.
Wave modelling has been stimulated by the Joint North Sea Wave
Project (JONSWAP), 1967-1975. The most recent activity in succession of JONSWAP has been the Sea Wave Modelling Project (SWAMP),
which has produced a comprehensive comparison of a large number
of numerical wave models; a report of this will be issued by
KNMI at the end of 1982.

2. NORTH SEA CLIMATE

Both authors of this paper are taking part in Working Group I -
Environment of the Conference on Safety and Pollution Safeguards
in the Development of North West European Offshore Mineral
Resources. A subgroup of this Working Group has written a report
which was submitted to the Conference at its session at Oslo,
May 1982. This report has been aimed primarily at controlling
designs of mobile installations. Such installations may be transferred regularly from one part of the North Sea to the other or
even from other parts of the World. It therefore seems more practical to define a set of standard design criteria which are valid
in all areas of the North Sea under the jurisdiction of the adjacent countries. The main subjects of the report are:

- wind speed, also including the gustiness of the wind;
- air- and sea temperatures, snow and ice (no serious problem in
  the North Sea);
- water depth, mean sea level, storm surges, tidal heights;
- waves and currents.

Seismicity is not covered by this report.
As to wind speed, generally this will not produce serious loads
on the whole body of a structure compared with e.g. waves;
however, waves and long-period water movements are generated by
the wind. Gusts may cause damage to members of installations.
Figures 1 and 2, taken from the report, show values with an average recurrence period of 50 years of hourly mean wind speed and
"3 second gust" speed (time length of gusts is matter of discussion among specialists in this field). Both figures show a
difference of about 20% between the northern and the southern
part of the North Sea. A factor of about 1.35 relates gust speed
and hourly mean wind speed for an observation height of 10 m
above the sea surface.
Figure 3 also taken from the report shows the 50-year storm wave
heights (crest-to-trough height of individual waves) for a

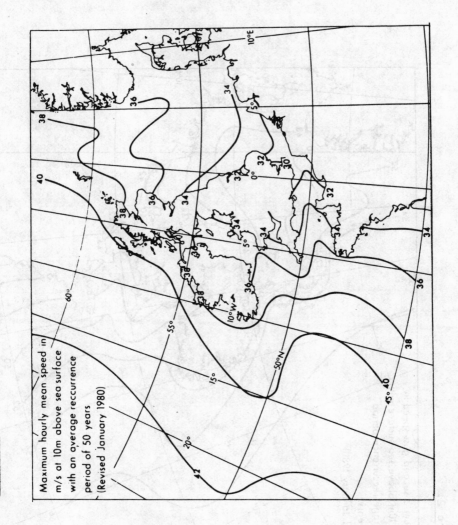

Figure 1.

Maximum hourly mean speed in m/s at 10m above sea surface with an average recurrence period of 50 years (Revised January 1980)

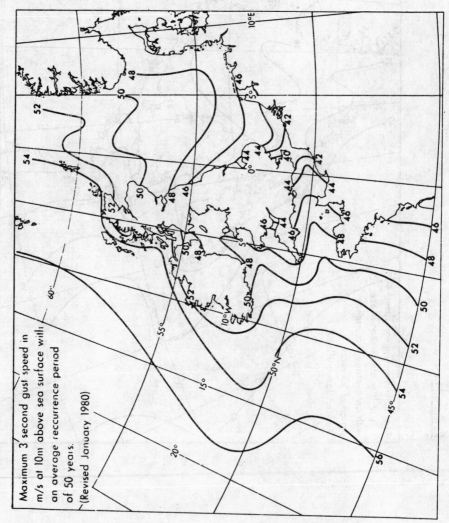

Figure 2.

Maximum 3 second gust speed in m/s at 10m above sea surface with an average reccurrence period of 50 years. (Revised January 1980)

Figure 3

Fifty-year storm wave heights for a fully-developed
storm lasting 12 hours.
Wave heights in metres.

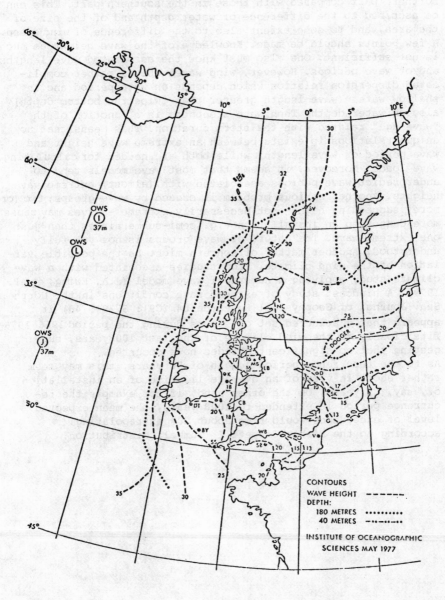

CONTOURS
WAVE HEIGHT   -------
DEPTH:
        180 METRES   ••••••••••
        40 METRES   -••-••-••-

INSTITUTE OF OCEANOGRAPHIC
SCIENCES MAY 1977

fully-developed storm lasting 12 hours. There may be slight
differences if another definition is used, but these are not
important here. We see that, unlike for wind speed, there is a
substantial difference between the extreme wave heights in the
northern part compared with those in the southern part. This can
be ascribed to the difference of water depth and of the size of
the area, and to some extent also to the difference of wind speed.
A few points should be made. Knowledge of the wave height as such
is not sufficient. One also must know the associated wave lengths
and/or wave periods. However, wind waves obey a rather complicated dispersion relation which depends on wave period and in
shallow water (wave length greater than twice the bottom depth)
also on water depth. Then wave steepness is a function of the
"wave-age" related with the storm duration. This means that no
unique relationship exists between an extreme wave height and
wave period or wave length, while both are needed for calculating
wave loads. Moreover, it seems that most wave models tend to
underpredict wave periods associated with (almost) extreme wave
heights. Another serious problem is caused by wave-groups; two (or
more) succeeding high - not necessarily extreme - waves may cause
more danger to an installation e.g. semi-submersibles than just
one extreme wave. The nature of wave groups is not yet fully
understood. Another matter of concern might be the possible variation of the wind climate. From studies associated with a wave
climate study applying a numerical wave model (J.A. Ewing, et al.,
1979). A hindcast study of extreme wave conditions in the North
Sea. Journal of Geophysical Research 84 (C9): 5739-5748) it
appears that a selected set of storms during the period 1966-1976
fits closely to the wind climate of the last 100 years, suggesting that no significant changes have occurred.
As to the recurrence period length of 50 years, this may seem
rather short in view of an average lifetime of an installation
of, say, 25 years. On the other hand, if for example the recurrence period is extended to 1000 years, the mean expected
level of exceedance would rise about 20% extrapolating
according to the Fisher-Tippett I (Gumbel) distribution.

SEAQUAKES, AN HAZARD TO OFFSHORE PLATFORMS?

K. Hove

Det Norske Veritas, Høvik, Norway

The present knowledge of seaquakes, or seismic acoustic shock waves in water, is scarce. Detailed studies on this phenomenon have not been undertaken. However, it is generally recognized that submerged structures or structural members that are not fully flooded will be subjected to large hydrodynamic pressure pulses from seismic acoustic shock waves in addition to the hydrostatic pressure. Less is known about the vibrational effects. However, several cases are reported in which ships in the open sea have experienced severe vibrations or have been damaged due to seaquakes. See Ref. 1.

Based on damage inflicted on ship structures that have been subjected to seaquakes it is to be expected that airfilled bracings on steel template structures will experience severe vibrations when subjected to strong aboustic shock waves.

In the epicentral region the effects of the acoustic shock waves, which are thought to be primarily the P-waves propagating through the water body, will occur simultaneously with the effects from the S-waves, and the two effects appear to be additive.

As regards damage to offshore platforms due to seaquakes the only offshore platform, to my knowledge, that has been subjected to a major earthquake is a steel template platform offshore China that was shaken by the 1976 Tangshan earthquake ($M \sim 8$, $d_e \sim 110$ km). In that case the structure itself survived. However, risers and equipment were torn loose and major damage was inflicted. Whether seaquake effects occurred in not known. Floating facilities are probably more vulnerable to seaquakes than fixed platforms.

A. R. Ritsema and A. Gürpınar (eds.), Seismicity and Seismic Risk in the Offshore North Sea Area, 149–150.
Copyright © 1983 by D. Reidel Publishing Company.

The characteristics of seaquakes are not well established.
Rational methods for considering seaquakes in the design of
structures and equipment systems remain to be developed.

REFERENCE

Hove, K., Selnes, P.B., Bungum H., 1982. Seaquakes. A Potential
      Threat to Offshore Structures. BOSS-82 Conference, MIT,
      Cambridge, USA, August 1982.

DISCUSSION-NOTES SECTION 3

Tides, Ocean Waves And Sea Level Changes

Q. A.R. Ritsema:

All types of tidal patterns do occur with one or more
clear or unclear maxima in a 24 hours record. In this
sequence there will be times that the gradient of the
wave motion is maximal. Has a significant correlation
been found between the occurrence of earthquakes and these
times of especially large tidal gradients?

A. P. Melchior:

This kind of changing pattern was a great concern for the
authors when they used the global gravimetric tide as a
"tidal clock": the distance to the nearest maximum was
difficult to determine when the inequality of the maxima
is growing. It is a reason why with the global method they
had problems at middle and high latitude where for gravimetric tides the inequality is very large.
This problem has disappeared since they used pure tidal
frequencies as clocks.
However, it is very interesting to see if larger gradients
may produce a larger number of earthquakes. For that purpose one has to look into the origin of the irregular tidal
pattern.
As the diurnal waves are proportional to the $\sin 2\delta$ and
the semidiurnal to $\cos^2 \delta$ of the concerned celestial body
(Sun or Moon) it is easily seen that such a pattern will
appear when $|\delta|$ is maximum and that the diurnal waves
vanish twice during each celestial revolution of the body.
The corresponding period is exactly Mf (13.66 d) and thus
any effect linked to this kind of pattern will appear with
the frequency of this wave.
The statistical analysis of the earthquakes for the Alaska-
Aleutian (region number 1) shows indeed a very significant
correlation with Mf.

Q. A.R. Ritsema:

Is the evidence for correlation between earthquake events
and tidal phase more clear for seismic zones with predominantly dip-slip fault motions than for those of the strike-
slip kind?

A. R. Ritsema and A. Gürpınar (eds.), Seismicity and Seismic Risk in the Offshore North Sea Area, 151–152.
Copyright © 1983 by D. Reidel Publishing Company.

A.   P. Melchior:
     So far the study made is purely statistical. The authors
     say that they intend to determine zones where the tidal
     correlation is high and later to look at the predominant
     mechanism in the area.

Q.   E. Booth:
     Levels of seismicity of certain parts of the world have
     varied markedly over the past 100-200 years. Is there any
     evidence of a variation in extreme values of wind speeds
     and wave heights in the North Sea over a similar period.

A.   E. Bouws:
     The amount of wave data is too limited to say anything on
     the variability of the wave climate. Visual observations
     sometimes are questionable and instrumental data are collec-
     ted only since around 1960. Moreover, variations of the
     wave climate are directly related to variations of the
     wind climate. Although some authors have observed local
     short-term variability of the wind speed (10-30 years) it
     seems that, generally speaking, no significant changes
     have occurred (cf. additional remark on this matter at
     the end of our paper).

Section 4

Instrumentation

INSTRUMENTATION FOR NORTH SEA SEISMIC DATA ACQUISITION *

T Turbitt, C W A Browitt, S N Morgan, R Newmark & D L Petrie
Institute of Geological Sciences, Edinburgh, UK

* Presented by Dr C W A Browitt

ABSTRACT

In an endeavour to improve the detection threshold and accuracy
of parameter determination for North Sea earthquakes, the IGS
and UK Department of Energy embarked on a programme of onshore
and offshore seismograph installations in 1979. On land, this
amounted to a geographical extension of the limited UK network,
particularly into northern Scotland and Shetland. At sea, new
techniques have been developed in order to deploy and maintain
a continuously recording system with sea-bed sensors linked via
a spar buoy to Mobil's production platform, Beryl Alpha. The
philosophy and technical details of the offshore station are
presented together with the principal difficulties and results.

INTRODUCTION

North Sea seismicity and seismic hazard have been inadequately
studied owing to the lack of historical macroseismic information
(inevitable for an area with zero population) and poor seismograph network geometry in recent times. The difficulties of
acquiring the fundamental data have been exacerbated because the
area is bordered by several countries each with a tendency to
look inwards to the region where earthquakes can be located more
accurately by the National network.

The deficiencies of the North Sea earthquake data base have
now been generally recognised. The method by which improvements
would be achieved from the UK through data exchange and the
installation of onshore and offshore seismographs was outlined

155

by Browitt in 1978 [1]. The present paper describes the
instrumentation which has been developed or deployed by the
Institute of Geological Sciences in pursuit of the data base
improvement.

LAND-BASED SEISMOGRAPHS

The IGS has been using the Earth Data Geostore analogue tape
recorder as the basis of its permanent seismograph networks
since 1975. The recorder will accept 10 channels of seismic
data frequency modulated with ±40% deviation on a carrier of
676 Hz. Selectable tape speeds of 15/160, 15/320 or 15/640
inches per second give recording times of 3½, 7 or 14 days for
seismic frequencies up to 32, 16 or 8 Hz respectively. Two
time channels (for internal and external clocks) and two flutter
channels are provided. For recording a small number of channels
(up to 5) the instrument will operate in a shuttle mode in which
the tape automatically reverses at the end of one pass giving
twice the recording time for the selected tape speed and bandwidth.

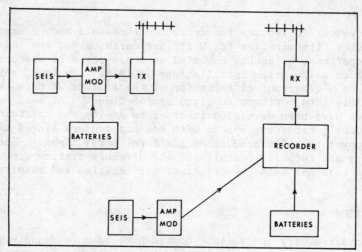

Figure 1. Schematic representation of a seismograph network
with land-line and UHF radio links to the recorder.

The typical network configuration is illustrated in Figure
1. In the normal mode of operation 6 or 7 vertical Willmore
Mark III short-period seismometers feed amplifier-modulator
units, with UHF radio transmission over distances up to about
100 km. An outstation will operate for one year with 12
disposable primary cells. Current drains are about 6ma for
the amplifier-modulator and 100ma for the radio transmitter.

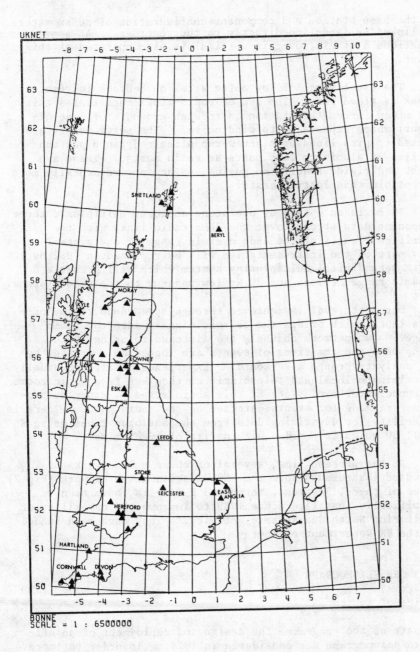

Figure 2. The UK seismograph network of the Institute of Geological Sciences.

At the base station a 3 component configuration of seismometers is linked by land-line directly to the recorder.   At some locations a single low-gain vertical is also operated at this point.

The usual ideal of low noise sites on bedrock in remote areas is aimed for.   The use of radio-links facilitates this but often forces the station on to high ground (for line-of-sight) which exposes it to wind noise.   The seismometers are housed in pits lined with screw-top plastic drain pipes and utilise 3m aluminium lamp posts as radio masts.   These are light and rigid, require no guying and are environmentally more acceptable than lattice masts.

There is an option of operating analogue multiplexed three component data streams over the 3kHz radio links with the sacrifice of half of the 45db dynamic range of the system. This part of the instrumentation will be replaced in 1982 by the Earth Data digital multiplexing system which will preserve a dynamic range of 96db and 30 Hz low pass up to the record stage.

Data retrieval is achieved through speeded-up replay of data tapes which brings the seismic signals into the audio range.   Two network channels are listened to using stereo headphones and the times of events are logged on a printer by the analyst through a keyboard.   Experienced staff can usually discriminate local and teleseismic earthquakes from sonic booms and underwater explosions at this audio stage.   Paper seismo-grams are made for events detected on the audio pass.   There is a facility for digitising data from the analogue tapes using a PDP11/50 which is loaded with a file of event times.

At the present time, several of these sub-networks of standard instruments form the IGS UK seismograph network (Fig 2). Three of these, Shetland, Moray and E. Anglia, have been specifically installed since 1979 to improve the capability of monitoring North Sea seismic activity.   They have been funded by the UK Department of Energy.

OFFSHORE SEISMOGRAPH

Choice of system

As part of the programme the design and deployment of an off-shore seismograph was considered in 1979.   In order to improve the chances of early success in acquiring data in the extremely hostile environment of the North Sea, we adopted a philosophy of using as much well proven technology as possible.   The land-based seismographs operate with few breakdowns and we, therefore,

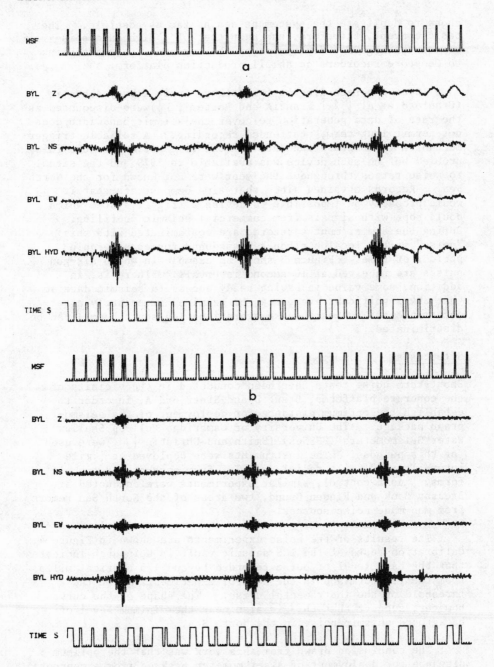

Figure 3. Air gun signals recorded on the vertical, north-south
and east-west seismometers and the hydrophone near Beryl Alpha.
a. filtered 0.005 to 15 Hz.   b. filtered 3 to 15 Hz.

chose to duplicate the system at sea as far as possible.   The
basic concept, therefore, has been that of a high gain seismometer package for deployment on the sea-bed with a data-link
to Geostore recorders on an oil production platform.

Ideas for recording at the package or in a buoy above it
(Cranford et al [2] Francis and Porter [3])were discounted as
the rate of data generation to cover the seismic bandwidth does
not permit long-term, unattended recording.   A reliable trigger
mechanism to compress the data in real time would overcome this
problem but no such device was available in 1979 and the signal
to noise ratios throughout the year were not known for the North
Sea.   Records obtained since that time demonstrate that it
would probably be impossible to devise a trigger system which
could cope with signals from commercial seismic profiling.
During the summer months records are contaminated with this
"noise" in the local earthquake pass-band for considerable
periods of time.   Figure 3 shows an example in which air-gun
pulses are received at 10 second intervals.   There is, in
addition, some value in having ready access to seismic data so
that any significant events can be examined immediately after
occurrence and non-earthquake vibrations (eg sonic booms) can be
discriminated.

Site tests

Short-term noise tests have been conducted in the vicinity of
the concrete platforms, Brent D and Statfjord A, in order to
establish the optimum distance for deployment of the seismograph package.   The University of Cambridge Pull-Up Shallow
Water Seismometers (PUSSes) (Smith and Christie [4])were used
for this purpose.   The instruments were deployed at $\frac{1}{4}$ mile
intervals on $2\frac{1}{2}$ mile radial profiles from each of the two platforms.   As a control, similar experiments were conducted at
Bressay Bank and Fladen Grund, two areas of the North Sea remote
from man-made noise sources.

The results of the noise experiments are shown in Figure 4.
Calibration tests at the IGS seismic vault in Edinburgh indicate
that the base level of noise recorded beyond 1.5 nautical miles
from the platform and in the control experiments is at the
threshold of the instrumental noise.   The shape of the curves,
however, suggest that this is also near the minimum level of
background seismic noise in the North Sea.

The conclusion drawn from this work was that the optimum
distance for deployment of a seismometer package from a concrete
platform of the Condeep design is 1.1 to 1.5 nautical miles.
At this distance platform generated noise is significantly
attenuated and the position is sufficiently close to the

platform and its attendant standby vessel for some degree of "policing" and protection from shipping to be achieved.

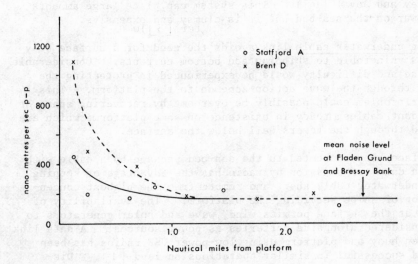

Figure 4. Background noise profiles obtained at Brent D, Statfjord A, Fladen Grund and Bressay Bank.

Data links

Three options have been considered for linking sea-bed sensors to recorders on oil platforms 1 to 1½ miles away.

An acoustic link is attractive in that it eliminates surface structures and sea-bottom cables which present a considerable target for fishing trawls and ship's anchors.   It was discounted, however, for the following reasons:

i)   The attachment of a receiver to a platform leg would be both technically difficult and expensive.

ii) They are prone to noise interference, particulary around offshore structures.

iii) They are demanding on power.

The possibility of using the RIPPLE range of isotope-powered thermoelectric generators (Myatt et al [5]) was considered but its power output was insufficient and there may be politico-

environmental problems in the deployment of a radio-active
source.   We learned of research into sea-water electrolyte cells
but these do not appear to be in production.   Lead-acid
batteries have been used underwater in oil-filled containers
(Higley and Joyal [6]).   This system can place large amounts
of power on the sea-bed but it is clumsy and expensive.

A long underwater cable link avoids the need for a surface buoy
but is vulnerable to shipping and bottom currents.   Considerable
expense and difficulty would be experienced in protecting the
cable through the wave action zone on to the platform.   This
latter problem could possibly be overcome by recovering spare or
redundant cables already in existence on some platforms which are
ducted through the towers well below the surface.

A surface buoy connected to the sea-bed package by a cable with
onward data transmission by radio has the advantage of keeping
the underwater cable short and remote from the highest concentration of ships servicing the platform.   The availability of
space at the surface permits wind, wave and solar generators to
be considered alongside batteries as power sources.   A data link
between buoy and platform using low-power UHF radios has been
proved successful in similar operations on land [1].   Disadvantages arise from the buoy being necessarily stationed in
the severe environment of the air-sea interface in the middle of
the North Sea.   Discussions with offshore operators have revealed that buoys are readily lost in stormy weather or in collision with shipping.

        This last data link involving a short cable, buoy and UHF
radio has been adopted.

Data buoy

The choice of a suitable buoy reduces broadly to a consideration
of size (and expense) and the type of mooring.   Many oceanographic systems have used surface following or wave-rider buoys
(Sutton et al [7], Rusby et al [8]).   Some of the disadvantages of these types are unpredictable loadings on the catenary
moorings, excessive working/wear on the underwater cable and
mooring points owing to the relatively large excursions of the
buoy, and the difficulty in laying a three-point anchor system.
The multi-anchor system is necessary to prevent buoy rotation
and the consequent "winding up" of the sea-bottom electrical
cable.   It can theoretically be overcome by using an inductive
swivel in a single-point mooring which incorporates both
mechanical and electrical members.   The IGS requirement was to
pass AC electrical power in one direction (to the package) and
4 FM signals in the other through the inductor.   Although under
development, no such device was proven in 1979 when the choice

of buoy was made.

The alternative to a wave-rider is a tension leg system
in which a spar buoy is held down in the water by a single taut
mooring attached to a heavy gravity base. Spinning is prevented
by a subsidiary catenary mooring with an anchor deployed to one
side. This system has the advantage of known loadings on the
moorings and a very stable configuration which minimises wear
at the mooring points and in the cable.

In order to avoid the problem of vandalism or theft, which
is not uncommon for unattended buoys, a relatively large one
was envisaged.

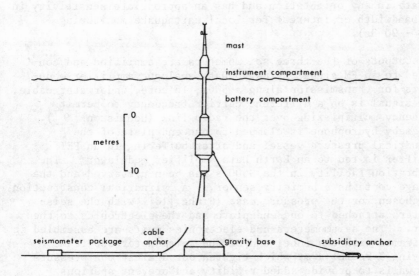

Figure 5. Sketch of the spar buoy seismic data acquisition
system deployed near Beryl Alpha in a water depth of 120m.

Havron Engineering Ltd designed and fabricated a spar buoy
to meet the required specification. It is 30 metres long with
a mast extending 10m above mean sea-level and sealed buoyancy
tanks at a depth of about 12m (see Figure 5). The slender top
section not only improves the dynamics of the buoy in response
to wave motion but also presents a small target and low inertia
for ship collision. The depth of the main buoyancy prevents
sinking after holing in a collision. The buoyancy of about
20 tons is held by the concrete gravity base weighing 45 tons in
water (about 80 tons in air). Cable entry at the bottom of

the buoy, 20m below the sea surface, ensures that it is protected
from the worst of the storm-driven wave motion,

Tank and computer model tests of the buoy motion predict
the maximum tilt of the system to be about 8⁰ in response to an
18m, 16 sec period wave.   This is equivalent to a watch circle
of approximately 30m in diameter.

Sea-bed instrumentation

The over-riding design criteria have been simplicity and proven
instruments wherever possible.   Sea-bottom systems have
previously utilised gimballed seismometers (Francis and Porter
[3], Smith and Christie [4]).   In order to avoid this
mechanical complexity the Geotech S500 was selected as it will
operate in any orientation and has an appropriate sensitivity in
the bandwidth of interest for local earthquake monitoring
(0.5 - 20 Hz).

Outputs of the three seismometers are amplified and converted to an FM signal by Earth Data analogue amplifier modu-
lators for transmission along a 500m, 16 core, underwater cable.
Each signal is on a different carrier frequency to permit
frequency multiplexing over the radio link (Houliston [9]).
A Graseby hydrophone is clamped to one end plate of the
cylindrical pressure vessel and after buffering in an FET
amplifier is fed to an Earth Data amplifier modulator.   The
calibration facility in the S500's has been preserved and the
package contains a humidity sensor.   A cylindrical construction
was chosen for the pressure case (Sinha [10]) with the seismometers attached to one end-plate and the electronics to the
other.   The seismometers and electronics units are assembled in
sub-frames with the inner end of the seismometer sub-frame
carrying 3 radial arms which tighten out against the pressure
case walls to provide added rigidity and prevent spurious
resonances.

The internal dimensions of the case are 100cm long by 25cm
diameter and the wall thickness is 0.8cm.   The package, with
the instruments, weighs 200kg in air and about 125kg in water.
It is bolted to a 50kg steel cradle fabricated from angle iron
to provide stability and improve coupling to the sandy bottom
(Avedik and Renard [11], Cranford et al [2]).

Materials such as bronze, stainless steel and marine
aluminium were considered for the pressure case.   However, cost
and availability favoured mild steel which does not corrode at
the installation depth (120m) owing to the lack of free oxygen.
This was empirically verified in May 1982 on recovery of a
package after a one year immersion at the Beryl seismograph

station.   Dissimilar metals have been avoided wherever possible
to prevent electrolytic corrosion.   Where this has not been
possible direct contact has been avoided by insulating with
polyacetal bushes.

Previous workers (Sutton et al [7]) found that 200m was a
sufficient distance between buoy anchor and package to escape
noise from buoy motions.   The selected cable allows for a 300m
separation and incorporates a small cable anchor on the sea-bed
between the catenary riser and the sea-bed run.

Prior to deployment the complete seismometer packages are
tested for leaks by lowering to 150m in a flooded mine shaft.

Buoy instrumentation

The buoy contains two UHF radio transmitters each capable of
handling three frequency-multiplexed seismic data channels on a
3kHZ bandwidth.   The underwater cable is ducted through the buoy
and is terminated in a distribution box where the four seismic
channels are summed for the transmitters.   One of the spare
radio channels is used for a battery voltage monitor.

Power for the instruments (about 300ma at 12 volts) and the
navigation light is supplied by eight packs of carbon-zinc
LeClanche cells supplied by AGA Navigation Aids Ltd.   These are
suspended in a 9 metre string down the narrow top section of the
buoy.   The stated capacity of 450AH means that the eight should
power the system for 15 months.

Two marine dipole antennae are mounted at the top of the
7m tripod mast and are connected to the instruments by cables
cast into steel-armoured hydraulic hose secured to the inner
side of the mast and led through the hull via Hawke glands.

The instrument and battery compartments are sealed-off
from the main buoyancy tanks below and can become flooded without affecting the integrity of the structure (this has in fact
happened).   Water is evacuated using compressed air which is
fed from divers' bottles through a Schrader valve in the hatch.
A narrow bilge pipe extends from the base of the battery chamber
to deck level for water to be ejected.

Platform instrumentation

This is a standard configuration used for land-based stations
and described previously.   It comprises two Geostore analogue
tape recorders, UHF radio receivers, demultiplexers and MSF
radio-time signal receivers housed on Mobil's Beryl Alpha platform.   Mobil's telecommunications staff change tapes every two

weeks and send them to IGS Edinburgh for analysis.   With this
tape-change interval a low pass of 16 Hz is achieved.

Deployment of the spar buoy system

An oil rig supply boat was used for deployment as these vessels
have the heavy duty winches necessary for the 80 ton gravity
base and the deck length to accommodate the buoy.

The gravity base was winched on to a steel cradle and held
by sea fastenings and the 52mm lowering wire attached to the
ship's main winch.   The main pendant was connected from the top
of the concrete block to the bottom of the buoy which was
secured along the ship's crash rail.

When on station the subsidiary anchor was lowered to the
seabed and its lowering wire left buoyed off.   The ship steamed
upwind whilst the buoy was lowered over the side and allowed to
stream astern.   The concrete base was lowered over the stern, a
smooth entry being achieved with the aid of the self-tipping
cradle.   As the base approached the seabed, the buoy was drawn
towards the stern by the main pendant and eventually assumed a
vertical attitude over the block.   The subsidiary anchor
prevented the buoy coming into contact with the stern.   It was
subsequently picked up and moved to the south to allow the
seismometer package to be deployed to the north.

A tripod mast with antennae, navigation light and connecting
cables was erected later.

The seabed instrumentation package was deployed at a line-
measured distance of 300m from the buoy using a small boat to
define the position.   Following the laying and transfer of the
electrical cable to the spar buoy, it was inserted through the
internal pipe from the base of buoy using divers.

Performance

The operational performance of the system is summarised below:

   i) July 1980:- spar buoy deployed.

  ii) October 1980:- seismometer package and buoy instrumen-
      tation installed.   Owing to adverse weather conditions
      only part of the battery complement was loaded.

 iii) December 1980:- batteries exhausted and data flow ceased.

  iv) February 1981:- buoy holed in a ship collision, battery
      chamber flooded and package and electrical cable lost.

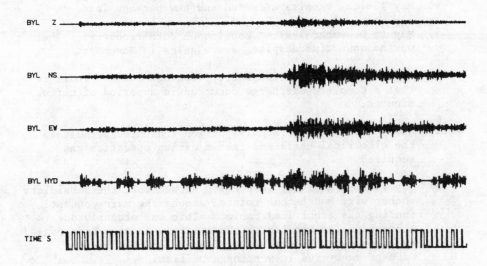

Figure 6. Beryl seismometer and hydrophone records of the
mb = 4.8 Skaggerak earthquake of 6 September 1981.

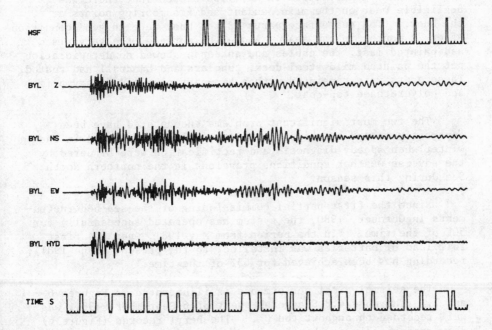

Figure 7. Beryl seismometer and hydrophone records of a small
magnitude local earthquake on 18 May 1982.

v)   May 1981:- repairs effected and new package laid.

vi) May to December 1981:- continuous data recovery. There
     was no down time despite severe gales in November.

vii) 21 December 1981:- the battery voltage monitor indicated
     that a complete discharge occurred in a period of a few
     minutes.

viii) March 1982:- a surface inspection indicated a fault in
     the electrical cable and that a diving operation was
     required.

ix) May 1982:- a diving inspection showed that the subsidiary
     anchor wire had become twisted around the main pendant
     fouling the electrical cable. This was probably due to
     a break on the sea-bed which is likely to have been caused
     by shipping. Repairs were effected and the sea-bed
     package recovered in working condition.

x)   May through June 1982:- continuous data recovery.

     The diving inspection in May 1982 showed that there was
negligible wear on the main pendant and its mooring points at
the buoy. The subsurface parts of the buoy were in good condition
after 22 months in the sea. Above the surface, the galvanised
and painted mast, the cables and antennae showed no deterioration
but the painted mild steel deck, tubulars and handrail had rusted.
These areas were repainted with a two-part aluminium epoxy primer
and polyurethane top-coat.

     The two most significant problems encountered have been
apparently caused by shipping. Both incidents occurred in the
winter when speedy diagnosis and rectification were hindered by
the adverse weather conditions prevalent in the northern North
Sea during this season.

     Since the first partial commissioning of the sea-bed instru-
ments in October, 1980, the system has operated successfully for
50% of the time. In the period from May 1981, when the first
full load of batteries was installed, to the present (June 1981),
recording has been achieved for 65% of the time.

     The seismological performance is illustrated in Figures 6
and 7. The mb = 4.8 Skaggerak event of 6 September, 1981, was
well recorded throughout the UK. The Beryl records (Figure 6)
show good signals on the three seismometers but the hydrophone
was partially saturated due to the swell. The 1982 package
has a reduced gain on this channel thereby improving the dynamic
range. Figure 7 shows the Beryl record of a small magnitude

local event which occurred on 18 May, 1982.   It was not detected
on land.

Two obvious sources of man-made noise have been identified
on the records (Figure 3).   In addition to the air gun signals
from seismic profiling work described previously, the unfiltered
record shows longer period noise (1.5 seconds).   This is the
same as the fundamental mode of oscillation of the concrete
Condeep platforms and Beryl Alpha is presumably the source.

FUTURE DEVELOPMENTS

The system has been demonstrated to work and it is, therefore,
tempting not to change it.   We intend, however, to make modifications to increase the dynamic range.   A seismometer package
will be constructed which utilises Earth Data three-component
digital multiplexing amplifiers.   These will provide a dynamic
range of 96db at the recorder from which a suitable "gain" can
be selected for the 45db available on the Geostore.   In parallel,
or subsequently, a digital recorder will be used in conjunction
with a microprocessor-based event triggering system if it proves
possible to effectively operate such a device with the range of
noise conditions in the North Sea.   The air gun profiling will
be particularly troublesome for a system based on an event trigger.

It is anticipated that by 1983 Mobil's accelerometer
channels from the base of the platform's utility shaft will become available for continuous recording on the IGS Geostore as
a strong motion system.

CONCLUSIONS

1.   New, land-based seismographs in Shetland and northern
Scotland have increased the detection capability for northern
North Sea earthquakes in the 1980's by a factor of about 5 over
the 1970's.

2.   The offshore seismograph near Beryl Alpha has operated
successfully.   The spar buoy hardware has survived for 23 months
in the period to June 1981 through 2 winters and a ship collision.
In the 13 months since May 1981 data recording has been achieved
for 65% of the time.

3.   The sea-bed seismometer package and electrical cable are
vulnerable to shipping activities and inspection and replacement
are hindered by adverse weather conditions in the winter.

4.   Commercial air gun profiling during the summer months has

been recognised as a significant source of noise which militates against unattended event-triggered instruments.

ACKNOWLEDGEMENTS

We are indebted to Mobil North Sea for logistic support on and around the Beryl Alpha production platform.   The work is supported by the Natural Environment Research Council and the Department of Energy and is published with the approval of the Director of the Institute of Geological Sciences (NERC).

REFERENCES

[1]   Browitt, C.W.A., 1979: Seismograph networks of the Institute of Geological Sciences, UK, Phys. Earth and Planet. Interiors, 18, pp. 127-134.

[2]   Cranford, M.D., Johnson, S.H., Bowers, J.E., McAlister, R.A. and Brown, B.T., 1976: A direct-recording ocean-bottom seismograph, Bull. seism. Soc. Am., 66, pp. 607-615.

[3]   Francis, T.J.G., Porter, I.T., Lane, R.D., Osborne, P.J., Pooley, J.E. and Tomkins, P.K., 1975: Ocean bottom seismograph, Marine Geophys. Res., 1, pp. 195-213.

[4]   Smith, W.A. and Christie, P.A.F., 1977: A Pull-Up Shallow Water Seismometer, Marine Geophys. Res., 3, pp. 235-250.

[5]   Myatt, J., Penn, A.W. and Poole, M.J., 1971: Isotope powered thermoelectric generators, Fourth United Nations International Conference on the Peaceful Uses of Atomic Energy, Geneva, 1971, A/CONF.49/P/487, pp. 1-7.

[6]   Higley, P.D. and Joyal, A.B., 1978: New mooring design for a telemetering offshore oceanographic buoy, Oceans '78, Washington DC, Marine Tech. Soc., pp. 10-17.

[7]   Sutton, G.H., Kasahara, J., Ichinose, W.N. and Byrne, D.A., 1977: Ocean bottom seismograph development at Hawaii Institute of Geophysics, Marine Geophys. Res., 3, pp. 153-177.

[8]   Rusby, J.S.M., Hunter, C.A., Kelley, R.F., Wall, J. and Butcher, J., 1978: The construction and offshore testing of the UK Data Buoy (DB1 Project), Oceanology International, 1978, pp. 64-80.

[9] Houliston, D.J., 1978: A frequency multiplexing
     transmission system, Inst. Geol. Sci., GSU Report No. 95.

[10] Sinha, M.C., 1978: Stresses and deformations in cylindrical
      pressure vessels, Dept. of Geodesy and Geophysics Univ.
      Cambridge Report, pp. 1-15.

[11] Avedik, F. and Renard, V., 1973: Seismic refraction on
      Continental shelves with detectors on sea floor, Geophys.
      Prospecting, 21, pp. 220-228.

STRONG MOTION EARTHQUAKE DATA ACQUISITION AND PROCESSING

Mustafa Ö. Erdik

Earthquake Engineering Research Center
Middle East Technical University, Ankara, Turkey

Earthquake strong motion data provide the basis for the design of critical structures and for the delineation of areas susceptible to ground failure as well as the basis for research on earthquake engineering and strong motion seismology. Consequently, the acquisition of the strong motion data is vital to the national and worldwide mitigation of the earthquake hazards.

This paper will discuss the status and recent developments in the strong ground motion data acquisition systems, indicate the areas where future developments are expected and provide and overview of the methods used in strong motion data processing.

RECORDING OF THE STRONG GROUND MOTION

Strong motion recorders are devices that record the time domain properties of the strong ground motion.

The maximum peak ground motions recorded are: 1,74 g for acceleration (1979 Imperial Valley Earthquake) and 1.16 m/s for velocity (1971 San Fernando Earthquake). Although the strong motion data for the ground displacement are highly unreliable, the maximum strong motion displacement can be presumed to be in the vicinity of 1 m.

Thus, in theory, a strong motion recorder should be capable of faithfully recording the ground motion from the ambient seismic noise levels to the levels somewhat above those recorded ground motion maxima.

The basic strong motion recorder utilized in the North and South America, South Asia and Europe is a triaxial, self-triggering analog accelerograph designed to record signals up to 1 g (recenty,

173

2 g) on 70 mm photographic film. The recorder has a dynamic range of about 55 dB and a frequency resolution of about 0.06-35 Hz. There exist about 5000 of such units installed worldwide.

Recently developed sophisticated digital strong motion recorders can reach dynamic ranges of 115 dB and frequency resolutions up to 300 Hz. A summary of the specifications of such digital recorders is provided in Table 1.

A strong motion recorder incorporates the following five distinct modules :

- Trigger
- Transducer
- Signal Conditioner (A/D Converter)
- Timer
- Recorder
- Power Supplier

Trigger: Triggers actuate the rest of the system when a preset treshold limit is exceeded. Triggers may be mechanical, in which a pendulum causes a switch closure when a preset displacement or tilt occurs, or electromagnetic, in which the coil voltage of a velocity transducer is amplified and compared to a preset value. The electro-magnetic triggers may be horizontal and/or vertical and typically have : band widths of 1-10 Hz, set points of 0.005-0.05 g and hold-on times of 6-20 sec. Suphisticated digital recorders utilize software based trigger algorithms with pre-event memories. The trigger algorithms are usually based on the ratio of the short-term-average (STA) of the incoming signal to the long-term-average (LTA) of the same signal. The duration of these averages and the threshold value of the ratio are specified for each case.

Transducer: The basic strong motion recording systems utilize an optical-mechanical acceleration transducer. However, various types of inductive, capacitive, piezo-electric and strain gage type transducers can be employed in strong motion data acquisition, systems.

For the measurement of acceleration; most of the commercially available digital recording systems incorporate the so-called Servo or Force-Balance Accelerometer type transducers. A Force-Balance Accelerometer incorporates a suspended mass with a capacitive type transducer attached to it. The displacement of the mass generates a servo-loop error signal, establishing a restoring force that is applied to the mass in such a way as to reduce the initial error signal to zero. The acceleration, which is proportional to the restoring force, is determined from the voltage drop across a resistance in series with a coil which generates the restoring force. A typical Force Balance Accelerometer has a flat response from DC to its natural frequency (usually 50 Hz) with a non linearity less than 1%. The transduc-

TABLE I. DIGITAL EVENT RECORDING SYSTEMS

	SPRENGNETHER DR-100	TERRA-TECHNOLOGY DCS-302	TELEDYNE MCR-600	KINEMETRICS PDR-2	ARGO/AS-280 GEOS-1
Pre-amplification	0-120 db gain in 6 db steps	X1, 5, 25, 100	6-120 db gain in 6 db steps	0-36 db gain in 6 db steps	0-60 db gain in 6 db steps
Number of Inputs	1-3	1-3 (6 optional)	1-3	1-3 (6 optional)	1-6
Dynamic Range	12 bit-72 db	12 bit+Gain Range 112 db	12 bit-72 db	12 bit+Gain Range 115 db	16 bit 96 db
Sample Rates	200 sps/ch 25 - 600 sps	100 sps/ch 50 - 600 sps	1-200 sps/ch	0.1-200 sps/ch	0.3-1200 sps/ch
Pre-event-memory	551-1670 samples	576 samples (192-15360)	1-999 samples	3072 samples	2560 samples
Trigger method	STA/LTA analog	STA/LTA digital	STA/LTA on filtered signal	STA/LTA modified with options	STA/LTA digital
Clock Stability	$\pm 5\times10^{-8}$ sec/day	5×10^{-6} sec/day	1×10^{-6} sec/day	$3-10^{-7}$ sec/day	$\pm 1\times10^{-7}$ sec/day
Recording Medium	Casette tape 2 tracks	Casette tape 2 tracks	Casette tape 2 tracks	Casette tape 4 tracks	Cartridge 4 tracks
Recording Time	4.6 min	7 min	6.5 min	27 min	35 min
Power Requirements Quiescent/Recording	$\pm$ 12 VDC 36/400 ma	$\pm$ 12 VDC 50/125 ma	$\pm$ 12 VDC 25 ma	$\pm$ 12 VDC 94/420 ma	$\pm$ 24 VDC 40/1.8 amp.
Operating temperature	0 to 70° C	-30 to 65° C	0 to 60° C	0 to 50° C	-20 to 60° C
Weight	13 kg	7 kg	13 kg	24 kg	16 kg

tion range is between 0.00001 to 2 g thus indicating a dynamic range of about 100 dB. The dynamic range of these transducers match to that of the digital recorders.

For the measurements of strong motion velocity; electro-magnetic strong motion velocity transducers are being utilized especially in Japan. These transducers are not generally available on a commercial basis. The typical transducer incorporated in the recently developed strong motion earthquake instruments array in Japan (Omote, 1980) has a natural frequency of 1-2 Hz and yield flat response between 0.05-50 Hz. These transducers have a measurement range of 0.0001-1 m/s with a corresponding dynamic range of 80 dB.

Modified long-period seismometers are employed for the measurement of earthquake ground displacement. The "PELS" transducers of the University of Tokyo (Omote, 1980) utilized in the recent Japanese Strong Motion Arrays have natural frequencies of 0.1-0.2 Hz with a 0.01-30 mm measurement and 60 dB dynamic range. It should be noted that the maximum measurement limit is far below that of the maximum strong motion displacement and these transducers need further development to be incorporated in reliable strong motion measurements.

Signal Conditioner and A/D Converter: The basic strong motion recorders do not include any provision for signal conditioning.

Digital recording systems generally include pre-amplification, antialias filtering and/or high-and low-pass filtering processes on the incoming signal. The incoming analog signal is converted into digital form with different number of bits, rain ranging and sample rates. The 16 bit and 12 bit A/D systems provide for linear dynamic ranges 96 dB and 72 dB, respectively. In digital systems with automatic gain ranging the microprocessor will alter the gain depending on the level of the last sampled signals with respect to the current full scale level. In the A/D conversion of such systems three bit binary gain code for each channel is included in the 16 bit sample word along with a channel identifier bit providing a dynamic range of 115 dB.

Timer: The simple basic recorders include provisions for timing marks for each half second. Internal time code generators or radio recievers can optionally provide the day, hour, minute, and second coding for event identification.

Sophisticated digital systems have automatic clock correction capabilities with GOES Satellite Receivers to alleviate the problems associated with clock stability.

Recorder: The basic analog recorders utilize 70 mm photographic film as the recording medium. The three-channel signals are recorded with a velocity of 10 mm/sec and sensitivity of 18 mm/g.

Sophisticated digital systems can record 6 channel signals on
cassette tapes. Type of recording is with phase encoding up to
1667 bpi density. Recording format can be either serial or blocked.
Recording time can extend to 35 minutes with 600 sps.

Power Supplier: The majority of the Strong-Motion Recorders run on
±12 VDC supplied through internal gell cells. The internal cells
are constantly charged with either mains power or solar cells.

During quiescent periods the current drain is about 20-50 m A.
However, during recording it reaches to about 400 m A with upto
1-2 A surges at startup.

FUTURE DEVELOPMENT EXPECTATIONS

The following strong motion recording system development expecta-
tions have been expressed in the International Workshop on Strong-
Motion Earthquake Instrument Arrays, May 2-5, 1978, Honolulu,
Hawaii (Iwan, W.D.,1978).

Relative Displacement Measurements to record the relative displacement using a comparative system providing electronic conditioning of data from a dual set of accelerometers.

Point Angular Measurements to record the rotational motion at a
point in acceleration or velocity mode.

DEPLOYMENT OF STRONG MOTION RECORDING SYSTEMS

Strong motion recording systems can be deployed in the field in
different configurations to meet the specific needs of data
acquisition.

Regional Seismicity Applications: A network of strong motion
recorders, located in an orderly manner to allow triggering of
many instruments by an event, will provide data on the amplitude,
frequency content and the duration of different sites on the same
earthquake. The use digital recording systems with buffer memories
enables the capture of the p-wave arrivals, and thus provide data
on the hypocenter computations.

Dense Array Applications: To study the source mechanisms and wave
propagation characteristics of earthquakes several array configurations have been proposed depending on the source shapes (Iwan,
1978).

Three dimensional local effects arrays would provide engineering
information about the modification effect of the local geologic
and topographic conditions and the mechanisms of soil failures.

Structural Response Applications: Compared to the model testing
on shaking tables, harmonic forced vibration generation and ambient
excitation surves, the acquisition of full scale, medium-to-high
strain structural response data due to ground shaking is of

primary importance in the design of all critical structures.
Several jurisdictions have passed national or internationals laws
which call for instrumentation of important structures.

STRONG MOTION DATA PROCESSING PROCEDURES

Computational methods in earthquake engineering require the acceleration, velocity and displacement trace of the ground motion or
of the structural response and various response and power spectra.
In studies of the source parameters such as size of earthquake
dislocation surface, effective stress, rupture velocity and the
displacement require good accuracy in the both ends of the Fourier
amplitude spectrum of the ground acceleration.

The accelerations obtained directly from the strong motion recordings would yield erroneous velocity and displacement traces, and
hence spectra, upon integration, due to the low and high frequency
noise.

The usable frequency band of a digitized strong motion accelerogram is inherently restricted by the combined noise of transducing,
digitizing and processing, and thus for proper processing and
utilization of accelerograms, various corrections has to be incorporated and the limits of this band have to be assessed.

The processing methodologies so far developed may be classified
in three major groups: (Erdik, 1980).

The first method assumes that the baseline of the acceleration
trace is of the polynomial form. This technique basically adds
a time dependent second order polynomial to the acceleration with
the coefficients of the polynomial computed so as to minimize the
mean square of the velocity. However, with the use of the higher
order polynomials, the baseline would begin to approximate the
negative of the accelerogram, eventually resulting in zero
acceleration, indicating the non-conservative characteristics of
this technique.

The second method which has found wide applicability in the United
States is based on bandpass filtering in the time domain. The
flowchart of the specific procedure as developed in California
Institute of Technology (Trifunac, Udwadia and Brady, 1971;
Trifunac and Lee, 1973) is provided in Figure 2 for illustration.
Another time domain procedure proposed by Sunder (1980) utilizes
optimal finite impulse response linear phase differentiators for
the instrument correction and infinite impulse response nonlinear
phase elliptic filters for the bandpass filtering of the data.
Bandpass filtering in the frequency domain is commonly employed
in Japan (Kurata, et.al., 1980).

A flow chart of the currently employed strong motion data processing procedure at the Earthquake Engineering Research Center of
the Middle East Technical University is provided in Fig.1 of
purposes of illustration.

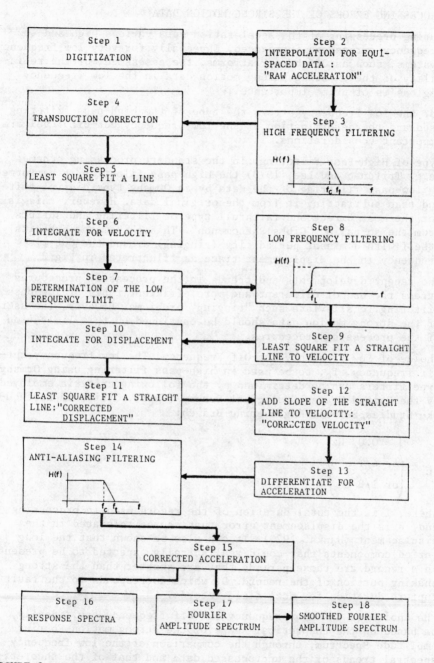

FIGURE 1. FLOWCHART FOR THE PROCESSING OF STRONG MOTION
ACCELEROGRAMS (Kubin and Erdik, 1981)

PROCESSING ERRORS OF THE STRONG MOTION DATA

During processing of the acceleration data various high and low
frequency errors are introduced. Especially for very low frequency
systems, such as Offshore Platforms, the assessment of the reliability of the existing strong motion data in the low frequency
regions is of prime importance.

For the low frequency errors the type of the high-pass filtering
used and the determination of the low frequency cut-off level are
important considerations.

Type of High-Pass Filtering: In the standart processing schemes
(e.g. Trifunac and Lee, 1973) the high-pass filtering is performed
by low-pass filtering of the data by an Ormsby type digital filter
and than subtracting it from the original data. However, this is
not the most direct and faithfull type of filtering and suffers
from the so-called "Gibbs Phenomenon". This phenomenon exhibits
itself with dominant periodicity (ringing) around the cut-off
frequency in the displacement trace as illustrated in Fig.2.

The recent developments and trends in the processing procedures
stress the use of different and more efficient types of low-pass
filtering to alleviate such "ringing" problems. However, the users
of the strong motion data should be cautioned on this limitation
of the processing procedures.

Choice of Low Frequency Cut-Off Frequency: The low frequency cut-
off frequency, f_c, to be used in high-pass filtering using Ormsby
type filters can be determined by the following criteria employed
by the United States Geological Survey (USGS), Office of Earthquake Studies, Seismic Engineering Branch:

$$f_c = \begin{Bmatrix} 0.07 \text{ Hz} \\ \text{or } 4/T \\ \text{or } 1/8 \text{ e} \end{Bmatrix}$$

where, T is the total duration of the record used in processing
and, e is the displacement error that can be tolerated in the
displacement (Hanks, 1975). It can also be shown that the long
period components that could realistically expected to be present
in a record are those periods that are shorter than the strong
shaking portion of the record. D, which corresponds to the fault
rupture duration for near field records.

The choice of the low frequency cut-off frequency can, ideally,
be based on the theoretical shape of the strong motion Fourier
Amplitude Spectrum, through the comparison of the low frequency
spectral trends of the unprocessed data and that of the theoretical trend and the assessment of the frequency where deviations
start (Kubin and Erdik, 1981).

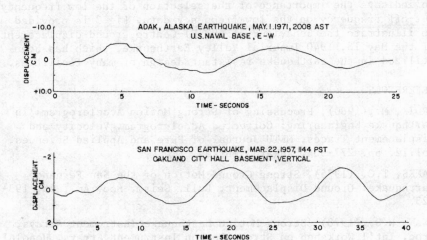

FIGURE 2. EFFECT OF RINGING ON CORRECTED GROUND DISPLACEMENTS

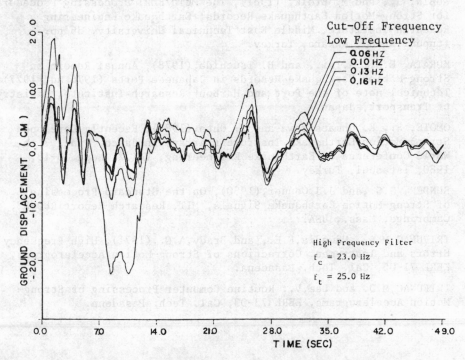

FIGURE 3. CORRECTED GROUND DISPLACEMENT TRACES FOR VARIOUS LOW
FREQUENCY CUT-OFF LIMITS (Imperial Valley Earthquake –
May 18, 1940-2037 PST El Centro Site Imperial Valley
Irrigation District, Comp: S00E)

To indicate the importance of the selection of the low frequency
cut-off frequency on the low frequency data, Fig.3 is provided
to illustrate the behaviour of the El-Centro ground displacement
of the May 18, 1940 Imperial Valley Earthquake, which has been
utilized in the earthquake resistant design of many facilities.

REFERENCES

ERDİK, M.,(1980), Processing of Strong Motion Accelerograms in
Earthquake Engineering: Corrected Accelerogram, Velocity and
Displacement Traces, METU Journal of Pure and Applied Sciences,
Vol.12, pp. 277-304, Ankara, Turkey

HANKS, T.C.,(1975), Strong Ground Motion of the San Fernando
Earthquake: Ground Displacement, Bull. Seism. Soc. Am., 65, 193-
225.

IWAN, W.D.,(1978), Strong-Motion Earthquake Instrument Arrays,
Proc. Int'l Workshop on Strong-Motion Instrument Arrays, Honolulu,
Hawaii.

KUBİN, J., and M. Erdik, (1981), The METU/EERI Processing Procedure
for Strong-Motion Earthquake Records; Earthquake Engineering
Research Institute, Middle East Technical University, 36 pp.
(unpublished), Ankara, Turkey.

KURATA, E.S., S.Iai., and H. Tsuchida,(1978), Annual Report on
Strong-Motion Earthquake Records in Japanese Ports (1976 and 1977)
Technical note of the Port and Harbour Research Institute, Ministry
of Transport, Japan.

OMOTE, S., K. Ohmatsuzawa and T. Ohta,(1980), Recently Developed
Strong-Motion Earthquake Instrument Arrays in Japan, Proc. 7 th
World Conference on Earthquake Engineering, Vol.2, Sept. 8-13,
1980, İstanbul, Turkey.

SUNDER, S.S.,and J.J.Conner,(1980), On the Standard Processing
of Strong-Motion Earthquake Signals, MIT, Research Report R80-83,
Cambridge, Mass., USA.

TRIFUNAC,M.D., Udwadia,F.E., and Brady,A.G.,(1971), High Frequency
Errors and Instrument Corrections of Strong-Motion Accelerograms,
EERL 71-05, Cal. Tech. Pasadena.

TRIFUNAC,M.D.,and Lee,V.: Routine Computer Processing of Strong-
Motion Accelerograms, EERL 73-03, Cal. Tech. Pasadena.

OCEAN BOTTOM SEISMIC INSTRUMENTATION

H. Ewoldsen

Woodward-Clyde Consultants, London, England

Of particular interest to resource development companies operating in moderate to highly seismic offshore zones, is the level of seismic motions to which their offshore structures may be subjected during the life of the facility. Moderate to large earthquakes may be detected by shore based seismic networks, with general location and focal depth determined from multiple recordings at a number of stations. However, as the exploration and production of offshore resources moves from several tens to perhaps several hundreds of miles from land, the precision of location, based on shore based information, becomes insufficient for purposes of engineering design of structures. Thus, the last several decades have seen the development of seafloor seismic instrumentation, both wholly self-contained and systems which transmit information to the sea surface through wire cables.

Initially, sea floor seismometers were used in near-shore configurations, with data transmitted to the nearby shore through electrical signals along cable laid on the ocean bottom. Early systems tested off the coast of California in the late 1950's and early 1960's used conventional seismometers in waterproof housings, with signal conditioning and recording completed at the land station.

As distances offshore increased, development was undertaken on fully independent seismograph systems which would rest on the sea floor, record seismic events, and transmit data to the surface on command. Two methods of data transmission were attempted; transmittal of data acoustically through the seawater column, which proved to be infeasible, and through the release of waterproofed cassette tape capsules which would float to the surface

183

A. R. Ritsema and A. Gürpınar (eds.), Seismicity and Seismic Risk in the Offshore North Sea Area, 183–185.
Copyright © 1983 by D. Reidel Publishing Company.

and be recovered by a surface vessel. After considerable development work, the Saudi Coporation of Alburquerque, New Mexico was successful in developing a system which could record up to 18 seconds of strong motion and remain operational for periods of six months or more without power pack changes.

The limitations of the above systems, together with their cost, probably means that for the near future, sea-floor seismograph installations will depend on hard-wire transmission of data from bottom seismometers to nearby offshore structures, moored buoys, or other similar devices. Particularly in the case of multiple seafloor seismometers, in which the intent of the system is to provide event location, focal depth, and sense of motion for small magnitude earthquakes, a single high cost moored structure supporting five to eight seafloor seismometers connected by seafloor cables would appear to be the most feasible way of proceeding.

For systems which are emplaced in offshore locations, it would be desirable to emplace instrument packages which have the ability to record both teleseismic and strong motion data. A combination of perhaps six to eight sensitive instruments for event location, and a single strong motion recorder to capture acceleration data from moderate to large nearby events, could provide very useful data to structural designers as well as the scientific world.

Of particular concern at the present time is the lack of a standard seafloor system configuration, specifically in the seismometer type and packaging. Land based seismograph systems, because of their longer history, have evolved into a set of several standard instrument types whose response characteristics are well known. In addition, comparisons between instruments of different makes have been carried out on numerous occasions, so that data from a given instrument can be readily integrated into the worldwide data base.

A somewhat less desirable situation appears to hold at present for seafloor systems. As most systems are designed to simply set on the seafloor, rather than being dug in, a variety of seismometer cases have been developed to provide low bearing pressures for muds, ensure adequate bottom contact through spike systems, etc. The result has been a series of seismometer cases which may have the response characteristics of flat plates, inverted pendulums, tethered balls, or rigid standing rods. An interesting comparison of the characteristics of signals from a dozen of the more widely accepted seafloor systems was carried out in 1978/1979 for the U.S. Navy's Office of Naval Research. In this test, all seismograph units were placed in a test location in Puget Sound, Washington, USA, at a depth of 10 meters, underlain by 3 meters of soft mud. The characteristics of the test area were such that all seismometers had virtually identical bottom conditions. Over a period of several weeks, events were recorded from nearby and distant earthquakes, quarry blasts, and air-gun experiments. Analysis of the data revealed a wide range of spectral response to each individual event, with evidence of seismometer-seafloor interaction, cross-coupling of seismometer and case, and other phenomena. As the structural engineer designing an offshore structure uses the spectral characteristics of the "design basis earthquake" to set his design limits, and as the design basis earthquake will in the future become more dependent on offshore seismic data, the need for standardization of instrumentation and correlation with land based systems is apparent.

TESTING METHODS RELATED TO SOIL- AND STRUCTURE BEHAVIOUR UNDER
DYNAMIC LOADING

J.G. de Gijt and W.R. van Hooydonk

Fugro Consultants International b.v., Leidschendam,
The Netherlands.

INTRODUCTION

   The design of dynamically loaded foundations requires data
on loading conditions and the response of structure and supporting
soil.

   Dozens of techniques are available to collect this information,
a few field- and laboratory tests that the author's company is
involved in will be discussed briefly in this paper.

   The interest in the methods described herein has grown considerably during the last couple of years. Main reason is that
relatively recent developments of testing and analytical techniques
now allow optimization of the design of structures that are
subjected to dynamic loading.

DYNAMIC LOADING

   Dynamic loads acting on a structure or soil may have a perio-
dic-, a random- or an impact character. Each type of loading generates a load-specific response and has associated problem areas.
Whereas periodic loads may typically cause fatigue, random and
impact loads may case strength and/or deformation problems.

   Figure 1 summarizes the wide range of dynamic loads that a
structure/soil may experience. In most cases this type of information can be used in the design phase. In areas where little
factual data exist, additional measurements have to be taken. The
latter is usually done by meteorologists, seismologist and hydro-

187

A. R. Ritsema and A. Gürpinar (eds.), Seismicity and Seismic Risk in the Offshore North Sea Area, 187–194.
Copyright © 1983 by D. Reidel Publishing Company.

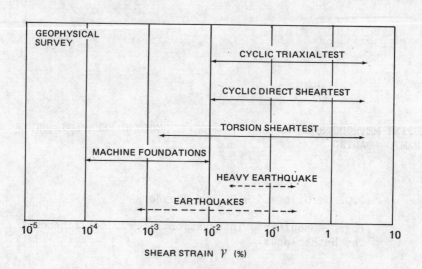

Fig. 1 Summary laboratory- and soil induced shear strain.

drologists. Wind-, current speed/direction, wave height, accelerations of soil masses during an earthquake are examples of
design-input. C.U.R. Report 57 (ref. 1) presents some of the
techniques used to gather the above data.

Mathematicians also play an important role when dynamic
loading occurs because statistical techniques and for example
Fourier-analyses are the basis for data reduction and interpretation. Von Karman and Biot (ref. 2) describe the most common
techniques for this type of work.

STRUCTURAL MONITORING

Although analytical methods often allow a resonable assessment of structural behaviour under loading uncertainties cannot
be avoided. For example the models for the superstructures and
the interaction between superstructure and foundation are subject
to discussion.

To be able to improve the models and/or check assumptions
made, monitoring takes place more and more. In its simplest form
acceleration measurements take place during (part of) the life of
a structure in combination with methods to measure the loads
activating the accelerometers. In addition, strains are sometimes
measured at crucial points if one is interested in fatigue data.

Dependent on the type of loading one can statistically determine the amount of measurements necessary to arrive at loading- and response spectra representative for the life of the structure. In case of permanent monitoring the monitoring-period may vary from six months to several years. Ansquer and Carton (ref. 3) describe an exercise like this.

If the information is required at shorter notice, forced vibrations are monitored. Multi-storey buildings can for example be excited by contra- rotating weights driven by a mini-computer. For offshore structures a mass of a few tonnes only, again controlled by a mini-computer, is sufficient to put it into motion. Any loading condition and subsequent response can thus be monitored relatively easy.

Another aspect of structural monitoring is the possibility to gather data on the integrity of parts of the structure. An interesting technique for members of an offshore-type steel structure is presented by Lepert (ref. 4).

SOILS UNDER DYNAMIC LOADING

When dynamically loaded, soils will exhibit strength and stress-strain characteristics that are completely different from those under static loading.
Type of soil, particle distribution, permeability and drainage conditions govern the behaviour of soil under loading. Wu (ref. 5) and ASTM (ref. 6) summarizes the phenomena and testing techniques for these type of problems. Dependent on the scope of work field and laboratory tests can be considered.

IN-SITU TESTS

To date a number of field testing techniques are available to assess parameters of soil under dynamic loading. On the other hand dynamic tests may also assist in determining static performance. The cross-hole test is an example of the first category, the dynamic pile test of the second. Both are described below.

Crosshole testing

The aim of this test is to determine the shear and Young's modulus of the soil. Essentially during the crosshole test one measures the travel time of shear and compression waves going from the source point in one borehole to the receiver points installed in, another, at least two other boreholes. As source of energy input for example a SPT-hammer can be used. This measurement can be repeated at intervals of about 1 to 3 m.

The result of the tests is displayed in a graph showing the travel time of the different waves. Knowing the distance between the points yields the velocity that is used to compute shear- and Young moduli based on elastic theory. Inaccuracies of the distance between source and receivers can easily yield deviations in computed shear- and Young moduli of approx. 20% for a difference in distance of about 5 – 10%.

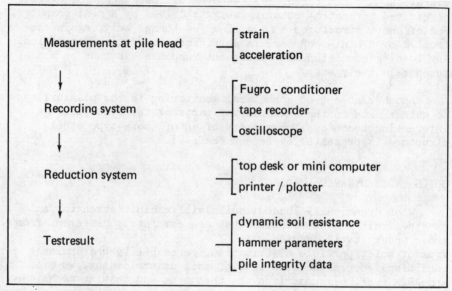

Fig. 2 Dynamic pile testing set-up.

Dynamic pile test

Another typical example of an in-situ test is a so called Dynamic Pile Test. Objective of this test is to determine, among others, the dynamic soil resistance that a pile meets during driving or, for cast-in-situ piles, during a blow after installation. The system is schematically presented in Figure 2. This dynamic resistance is of course subject to damping, changes of pore water pressure and stress conditions in the soil. Using soil-specific correlations, the dynamic resistance can be transferred into a static resistance, i.e. the pile capacity. This relatively simple test thus replaces time consuming and expensive load tests. Wright, Van Hooydonk and Pluimgraaff (ref. 8) describe such a test performed in clays.

LABORATORY TESTS

The existing laboratory testing techniques to investigate
the dynamic behaviour of soil are cyclic simple shear tests,
resonant column test, cyclic torsional test and cyclic triaxial
test. The latter test is discussed below.

Fig. 3 General cyclic triaxial testing set-up.

Cyclic Triaxial Test

In figures 3 and 4 a typical cyclic triaxial testing equipment is shown as placed in the Fugro B.V. Laboratory.
With this equipment a soil specimen can be subjected to a wide
variety of loading conditions and frequencies whereby the loads
on the specimens are measured electrically with an internal load
cell. Dependent on the initial stress conditions, isotropic
or anisotropic, a specimen is subjected to a number of cycles,
whereby the following items are continuously recorded:

- vertical deformation
- vertical load
- pore pressure
- cell pressure

The accurate measurement of the imposed load on the specimen
is illustrated in figure 5, where the results of a series of

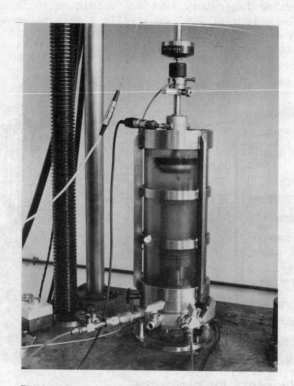

Fig. 4 High pressure triaxial cell for dynamic testing.

liquefaction tests is displayed.
From this figure it can be observed that the friction between
bushing and piston, if not known exactly, may overestimate the
liquefaction strength of the tested material substancially.
De Gijt and Beringen (ref. 9) describe the cyclic triaxial test
in detail.

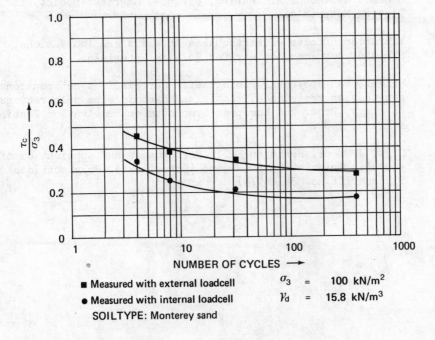

■ Measured with external loadcell σ_3 =   100 kN/m²
● Measured with internal loadcell γ_d =   15.8 kN/m³
   SOILTYPE: Monterey sand

Fig. 5 Comparison of internally- and externally measured force on
       liquefaction strength.

REFERENCES

1.   CUR-Report 57: Dynamic problems associated with civil
     engineering structures (in Dutch), Holland, 1972.

2.   Von Karman and Biot, Mathematical methods in engineering,
     McGraw-Hill Book Cy, New York, 1960.

3.   Ansquer and Carton, Field measurements of correlations
     between waves and platform response versus significant wave
     height and direction, Paper 3797, Offshore Technology
     Conference, Houston, Texas, 1980.

4.   Lepert, Vibro detection applied to offshore platforms,
     Paper 3918, Offshore Technology Conference, Houston, Texas,
     1980.

5.   Wu, Soil Dynamics, Allyn and Bacon Publ., Boston, Massa-
     chusetts, 1971.

6.   Dynamic Geotechnical Testing, STP 654, American Society for Testing Materials.

7.   Vibration effects of Earthquakes on soils and foundations, STP 450, American Society for Testing Materials.

8.   Wright, Van Hooydonk, Pluimgraaff: Correlations between cone penetration resistance, static and dynamic pipe pile response in clays, Proc. 2nd European Symposium on Penetration Testing, Holland, 1982.

9.   Gijt, J.G. de, Beringen F.L.: Performance and application of a cyclic Triaxial Test on sand (Dutch text). Polytechnisch Tijdschrift no. 34 (1979).

NOTE ON THE DETECTION AND LOCATION CAPABILITY OF THE NETHERLANDS
SEISMIC STATIONS FOR EARTHQUAKES IN THE NORTH SEA BASIN

A. Reinier Ritsema

Royal Netherlands Meteorological Institute,
De Bilt, The Netherlands

Six years of NEIS data lists have been searched for Netherlands station performance in the distance range of 0 - 10$^\circ\Delta$. The
lower m_b values detected at each station for certain distances
have been smoothed to the relations shown in figure 1.

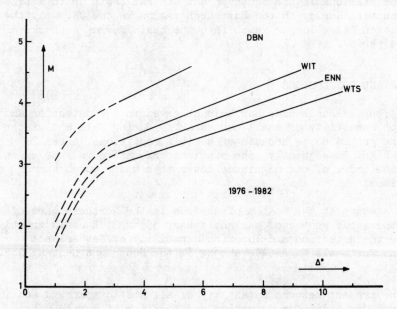

Figure 1. Detection capability of Netherlands stations for
near seismic events.

A. R. Ritsema and A. Gürpınar (eds.), Seismicity and Seismic Risk in the Offshore North Sea Area, 195–197.

The curve for the EPEN station (code ENN, formerly EPN), located on mid-Carboniferous quartzite, also includes data of the nearby HEERLEN station (HEE). In the later years the ENN station is performing at a better rate than before. It is now in many cases the better station than Winterswijk (WTS), located on mid-Triassic limestone. Station WITTEVEEN (WIT) is located at a noisier place on Pleistocene sands.

Stations WIT, WTS and ENN-HEE all are equipped with vertical MK-II Willmore sensors. Station De Bilt (DBN) has only middle- and long-period instruments of the Galitzin, Press-Ewing and Teledyne type.

The change in slope of the detection capability at a distance of about 2^O and magnitude $m_b = 3$ is probably caused by the relative inadequate generation of surface waves for the smaller magnitude events, together with a possible m_b bias because of the use of P_g instead of P_n waves in the case of short epicentral distances.

The result of this study for the North Sea basin is given in the figure 2. It shows that m_b 3 shocks only can be detected by three of the stations simultaneously in a small region outside the Netherlands coast. M_b 4 earthquakes can be detected and located by three stations in the Southern and Central basin up to a line from Southern Norway to the Edinburgh region in the UK. Only the WTS station has a longer reach into the basin for m_b 4 shocks in this period.

REMARKS AND CONCLUSIONS

Strong seasonal bias has been observed in the detection and location capability of small earthquakes. During fall and winter time the ground noise because of weather conditions, often is rather high. Subsequently, the station performance in such cases is of the order of one magnitude lower than under quiet summer conditions.

A station at the bottom of the sea in the Northern part of the Netherlands sector of the shelf near $55\frac{1}{2}^O$N $3\frac{1}{2}^O$E could greatly improve the detection and location capability of Netherlands stations for small magnitude shocks in the South and Central North Sea basin.

The present network capability of all stations around and in the North Sea basin for detection and location of events within the basin is being processed now.

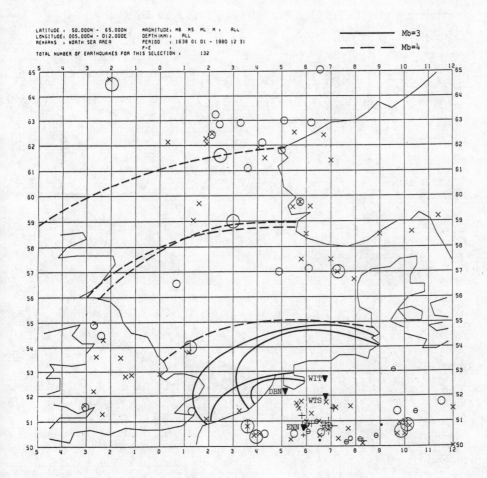

<u>Figure 2.</u> Detection and location capability of the Netherlands
          seismic network for earthquakes in the North Sea area.
          Full lines for m_b = 3, dashed lines for m_b = 4.

DISCUSSION-NOTES SECTION 4

Instrumentation

Q. O.T. Gudmestad:
Has Dr. Browitt had the possibility to summarize and correlate the findings of the recordings from the sea-bottom
seismometer with the recordings from land stations? In
particular, I am interested in the gain in accuracy of the
epicenter location, the earthquake depth and the attenuation determinations.

A. C.W.A. Browitt:
Many of the earthquakes recorded at the Beryl station are
also recorded on land. The greater the number of seismograms from each event and the nearer the stations are to
the epicentre, the better is the accuracy of location.

Q. O.T. Gudmestad:
I like to know which earthquake magnitudes can be detected
with the present North Sea bottom instruments which cannot be detected from the Shetland and the Bergen seismometers.

A. C.W.A. Browitt:
There have been several earthquakes in the past year which
have not been detected on Shetland and many more which
have not been detected at Bergen. The threshold of detection for both land-based and sea-bottom seismographs is
dependent on the background noise. This varies with the
weather conditions and, for Bergen, with the time of day
owing to the "cultural noise" of the town. Shetland frequently detects events greater than magnitude 2.5 ML from
mid North Sea areas. The detection threshold is, however,
greater than this for much of the winter season. It should
be noted that events near the threshold of detection cannot usually be well located owing to the inaccuracies in
timing onsets which are emergent from the background noise.

A. R. Ritsema and A. Gürpınar (eds.), Seismicity and Seismic Risk in the Offshore North Sea Area, 199.
Copyright © 1983 by D. Reidel Publishing Company.

SECTION 5

SOILMECHANICS, LIQUEFACTION, GEOTECHNOLOGY

SOME REMARKS ON THE CONTRIBUTION OF GEOTECHNICAL ENGINEERING
TO AN ESTIMATION OF SEISMIC RISK.

ir. H.L. Koning

Delft Soil Mechanics Laboratory
Delft, The Netherlands.

1.   INTRODUCTION

      In this paper an outline will be given of the contribution
which geotechnical engineering can give to an estimation of the
seismic risk of a site. Fig. 1. gives a very schematic picture of
some phenomena which can occur during an earthquake. The earthquake originates in the hypocenter and in relation with this the
first question can be posed: what is the magnitude of the earthquake to be expected?

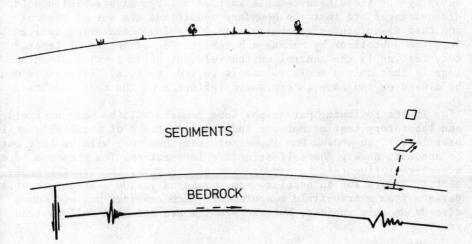

SEDIMENTS

BEDROCK

Fig. 1. What happens during an earthquake?

A. R. Ritsema and A. Gürpınar (eds.), Seismicity and Seismic Risk in the Offshore North Sea Area, 203–218.
Copyright © 1983 by D. Reidel Publishing Company.

Usually two kinds of earthquakes are considered:
- the operating base earthquake which has a certain probability of occurrence during the lifetime of the construction (dependent on the type of construction); due to this earthquake no permanent deformations of the construction are allowed to occur;

- the maximum credible earthquake which has a much lower probability of occurrence and which one may cause some permanent deformation, however, without affecting the overall safety of the structure.

Then the second question arises: what happens to the waves emanating from the hypocenter during propagating through the bedrock to the considered site? The answers to these two questions give the data with respect to the earthquake to be expected in the bedrock beneath the site. Important in this connection is the maximum acceleration, the frequency content and the duration of the earthquake.

Finally the waves propagate through the sedimentary layers to the surface and now the question is what kind of waves arrive at the surface. To answer this question the layering and the properties of these sedimentary layers must be known, particularly the strength, stiffness and damping properties are important. Obviously the co-operation of a geologist, a seismologist and a geotechnical engineer is very important to get the answers to the posed questions

The properties of the sedimentary layers can be determined in the field and/or in the laboratory by testing of samples. The advantage of field measurements is that a large mass is influenced; disadvantages are that the boundary conditions are out of control and that disturbances can be caused by pushing a measuring device into the subsoil or by making a borehole. The advantage of laboratory testing is the control on the relevant parameters; a disadvantage is that only a small volume is tested, so local differences can be missed or can have an excessive influence on the test results.

In the following paragraphs some remarks will be made on field and laboratory testing methods and the importance of a reliable soil profile will be shown. One important item, however, will be left out of account, namely the soil-structure interaction. The presence of a structure influences the propagation of the waves through the sedimentary layers and an accelerogram, measured inside a structure will deviate from a free-field measurement. Much research has been done already on this subject, but the problem has not been fully solved yet.

2.   FIELD INVESTIGATIONS

    These investigations can be subdivided into surface measure-
ments, measurements using a borehole, measurements using a push-in
device and combinations of these.

    As surface measurements refraction and reflection surveys can
be mentioned. In these methods waves are triggered on or at some
depth below the surface and are recorded by sensors at several distances placed on the surface. From the obtained results the layering of the sediments can be derived, as the velocity of wave propagation in these layers. These surveys must reach a great depth,
preferably to bed-rock, which often is taken as the layer in which
the velocity of the shearwaves amounts to from 600 to 900 m/sec.
Also a high resolution is required as thin layers can be very
important. Finally the determination of the velocities of the waves
must be very accurate, as small errors in these velocities cause
great errors in the properties of the layers derived from these
velocities.

    In up-hole and down-hole surveys use is made of a borehole. In
an up-hole survey waves are triggered successively at various depths
and recorded at the surface; just the reverse is done in a down-
hole survey. Two or three boreholes are used in a cross-hole survey.
In one of these boreholes waves are triggered successively at
various depths and recorded in the other borehole(s) at the same
depths. These measurements are possible only to the restricted depth
of the borehole(s). Apart from that the same remarks hold as made in
relation to the surface surveys.

    A combined method is the standard penetration test. In this
test a borehole is made to a certain depth and then a tube of prescribed shape and dimensions is in a prescribed way hammered into
the soil below the borehole. The number of blows required to reach
a certain penetration -the N-value- is empirically related to
several properties of the soil. This test is repeated at several
depths. However, the results depend heavily on the way of execution.

    A more objective method of field investigation is the cone-
penetration test. In this test a rod provided with a cone is pushed
into the subsoil with a constant velocity. The resistance which the
cone itself meets, is recorded and in this way a continuous profile
of the subsoil is obtained. Usually the rod is also provided with a
friction-sleeve just above the cone and so the local skin friction
can be measured continuously too. The ratio of the two mentioned
results is related to the soil type. The cone and friction resistances are correlated to several properties of the soil. These
measurements are possible to a restricted depth, depending on the
capacity of the apparatus used.

As will be shown in paragraph 3.3 the porosity of sand is a
very important parameter. For that reason the Delft Soil Mechanics
Laboratory developed a device to determine this porosity. Above the
cone and the friction-sleeve four electrodes are fitted around the
sounding rod. With the aid of this probe the specific electrical
resistivity of the soil is measured at every 20 cm difference in
depth. The specific electrical resistivity of the porewater is at
the same depths measured with the aid of a second probe. The ratio
of these two resistivities is related to the porosity; the relation
is determined in the laboratory on samples taken from the mass of
sand under investigation. It may be clear that this method is
applicable in saturated sandlayers only. The attainable depth is the
same as in case of a cone penetration test.

Summarizing: the most detailed information is supplied by an
up-hole, a down-hole or a cross-hole survey, by a borehole, or a
cone penetration test, the last one combined with a porosity
measurement if required. However, these methods all suffer from a
restricted attainable depth. A sufficient depth can be reached by a
reflection or a refraction survey; now difficulties can arise in a
many-layer subsoil. Obviously a combination of different types of
field investigations seems appropriate.

3.   LABORATORY INVESTIGATIONS

3.1 General

It goes without saying that in earthquake problems dynamic
testing of samples is required. Several apparatuses are available
for this purpose. Among these the cyclic triaxial apparatus is the
most frequently used one.

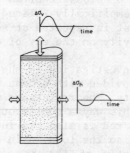

Fig. 2. Principle of the cyclic triaxial test.

In this apparatus, schematically shown in figure 2., a sample is
placed and an all-around pressure is executed (sometimes unequal
vertical and horizontal stresses are applied).

Then at least the Delft Soil Mechanics Laboratory prefers to do so,
the vertical and horizontal stresses are cyclically changed in such
a way that the increase (decrease) of the vertical stress is twice
the decrease (increase) of the horizontal stress. In this manner
the octahedral stress remains constant, which means that possible
changes in volume – especially important in case of sand – are
caused mainly by shear stresses. However, in most other labora-
tories the horizontal stress is kept at a constant level.

In figure 3. the simple shear apparatus is depicted. Figure 3a.
shows the sample in an undeformed state and figure 3b. in a deformed
state. This way of deformation has more resemblance with the defor-

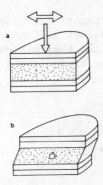

Figure 3. Principle of the simple shear test.

mations in the sedimentary layers when a horizontally polarized
shearwave goes up from the bedrock than the way of deformation in a
cyclic triaxial test. The state of stress in the simple shear test
is not as simple as in the triaxial test, however. Both in the triaxial test and in the simple shear test the shear-deformation
amounts to from 0.01 to 1%.

Lastly the resonant column apparatus must be mentioned. In
this apparatus (fig. 4.) a sample is placed and vertical and horizontal pressures are introduced (not necessarily equal to each
other). Then one end of the sample is cyclically excited, while the
other end remains fixed. Usually a torsional movement is executed,
so shearwaves are developed. In this apparatus shear-deformations
from 10^{-4} to 10^{-2}% are brought about then.

Two types of testing can be distinguished: tests aimed at
determining the strength characteristics and tests to establish the
deformation characteristics of the soil. Tests to determine the
strength of the soil are continued until failure of the sample is
observed. The number of cycles to cause failure is related to the
stress-level applied in the test and the physical characteristics of
the sample.

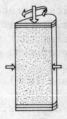

Fig. 4.   Principle of the resonant column test.

In tests executed to establish the deformational properties of
the soil the relation between the applied stress and the consequent-
ial deformations is recorded. In figure 5. the typical shape of an
obtained recording is drawn. From the slope of the chord of the

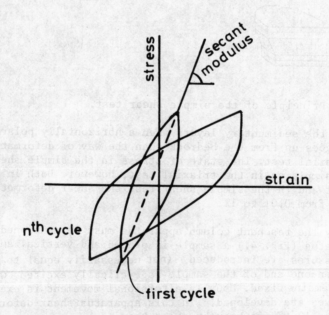

Fig. 5.   Typical result of a two-sided loading test.

figure the shear modulus or Young's modulus - dependent on the kind
of the applied stress - can be determined. The area of the figure is
related to the damping in the soil. The modulus decreases and the
damping increases as the number of cycles increases. Again these

results are related to the stress-level in the test and the physical characteristics of the sample.

In the following paragraphs will be gone into the results of laboratory investigations on clay and on sand. However, it must be stressed that the data given in no way are representative for all clay-soils or sands; their only purpose is to give a general view into the influence of several parameters.

3.2   Investigations on clay

Out of the results of many tests on clays-soils it turned out that the strength as well as the deformation characteristics are strongly dependent on the overconsolidation ratio (OCR). The OCR is the ratio of the highest vertical effective stress in the geological past to the vertical effective stress nowadays. This vertical stress can be diminished by erosion of surface layers or by melting of an icecover.

The number of cycles to failure decreases with an increasing overconsolidation ratio. As an example the results of simple shear tests on Drammen clay are shown in figure 6. [1]. In this figure

τ_{hc} =   cyclic shear stress,

τ_{hf} =   horizontal shear stress at failure for static loading,

γ_c =   cyclic shear strain.

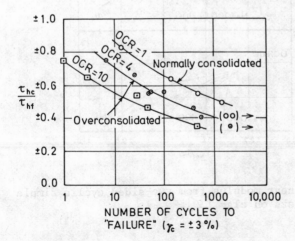

Fig. 6.   Number of cycles to cause failure in two-sided cyclic
          simple shear tests on clay   [1].

From this figure it also follows that the shear-stress level is
very important: the higher this level the lower the number of
cycles to failure or, the other way round, the strength decreases
with the number of cycles to failure.

    Also the shear modulus decreases as the OCR increases. In
figure 7. [2] the results are shown of simple shear tests on
Drammen clay. Once again the influence of the stress-level is
obvious: the shear modulus increases as the stress-level decreases.
Further it follows that the shear modulus decreases as the number of
cycles increases, as already was mentioned in the foregoing paragraph. However, at low stress-levels the modulus turns out to be

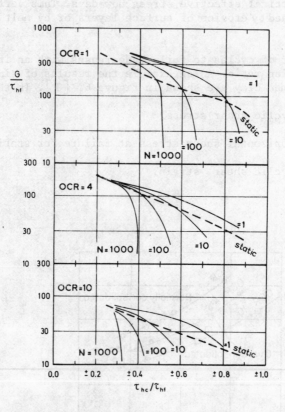

Fig. 7. Secant shear modulus from two-sided cyclic simple
         shear tests on clay [2].

independent of the number of cycles: then very small strains occur
and the clay-soil behaves approximately elastically.

3.3 Investigations on sand

Much research has been done on the strength characteristics
of sand. This is due to the fact that sand liquefaction phenomena
shows, which can have very dramatic consequences. It may be sufficient to recall the damage caused by sand liquefaction in
Anchorage, Alaska and in Niigata, Japan, following the Alaska and
Niigata earthquakes in 1964.

Consider a mass of loose sand, i.e. sand with a high porosity
(defined as the ratio of the volume of pores of a given mass of
sand to the total volume of that mass). When shear stresses are
exerted on such a mass of sand, a reduction in volume is caused.
If the pores are filled with water this reduction of volume must
result in the outflow of an equivalent volume of water. Excess
porewater pressures are created then. Through these excess porewater pressures the effective stresses between the individual sand-
grains are reduced. This increase in water-pressure may be so great
that these effective stresses become zero. Then the sand behaves
like a heavy fluid: liquefaction has occurred. In a cyclic test on
loose sand the excess porewater pressure gradually increases with
the number of cycles until liquefaction occurs. In the early stages
of the test, when the excess porewater pressures still have a restricted value, the deformations are very small. Towards the end of
the test the deformations grow and when liquefaction has been reached deformations of 10% and more are found.

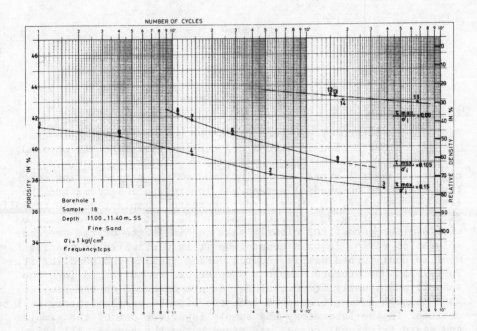

Figure 8.   Number of cycles to cause liquefaction in two-sided
            cyclic triaxial tests on sand.

In figure 8. the results of cyclic triaxial tests on fine sand
are presented. These tests have been executed at several porosities
and at several shear-stress levels on reconstituted samples; the
octahedral stress was kept constant at a value of 1 kgf/cm2 and the
frequency of loading was 1 cycle per second. From these results it
follows that both the porosity and the shear-stress level have a
definite influence on the number of cycles to cause liquefaction.

Again the overconsolidation ratio is an important parameter in
case of testing sand-samples. Figure 9. shows the results of some
cyclic triaxial tests on reconstituted, pre-loaded samples. The
horizontal axis gives the numbers of cycles to liquefaction and
the vertical axis the consolidation pressure. This last means that a

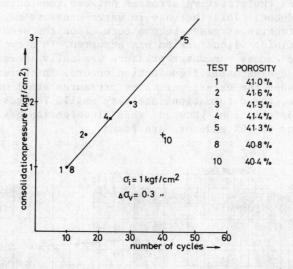

Fig. 9.   Influence of pre-loading on the number of cycles to cause
          liquefaction in two-sided cyclic triaxial tests on sand.

higher octahedral stress was maintained on the sample during some
time, then the octahedral stress was lowered to 1 kgf/cm2 and the
test was executed. In other words the tested samples had overconsolidation ratios ranging between 1 and 3. The test results on samples
with similar porosities clearly show the influence of the overconsolidation ratio. In this connection it is worthwhile to consider
what an undisturbed sample endures during taking and handling it.
To take an undisturbed sample a sampler has to be pushed into the
subsoil and usually frictional forces are executed on the sample.
What happens during transport from the site to the laboratory will be
omitted. In the laboratory the sample is driven out of the sampler
and again frictional forces are executed on the sample. All this

means that it is very well possible that a pre-loaded sample is
tested and consequently that favourable results have been obtained.
No doubt this conclusion also holds good for stiff clay-soils.

4.   INFLUENCE OF SOIL PROFILE

When the design-earthquake has been established and the soil-
profile and the properties of the different layers have been determined, then calculations can be made as to what happens at the
surface. In this paragraph the results of some calculations, made
with the aid of the computer program SHAKE [3], will be shown. In
these calculations an idealized accelerogram in the bedrock beneath
the site has been introduced. The chosen type of ground motion is
generated by slip over a prestressed penny-shaped crack in an elastic
solid [4] and was recorded during the Port Hueneme earthquake of
March 18, 1957 [5]. This accelerogram was selected because it permits a simple comparison with the response at the surface. In all
calculations the depth to bedrock amounted to 330 feet (100 m), the
maximum acceleration in the bedrock was supposed to be 0.1 g.

Figure 10. presents the results of the calculations in case the
sedimentary layers are composed of clay of which the properties increase with depth. At the right-hand side of the figure both the
maximum acceleration and the shear-stress levels are given as function of the depth.

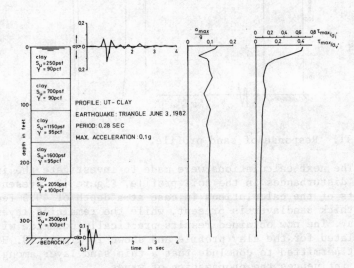

Fig. 10.   Response of clay profile.

At the top of the figure the accelerogram, calculated for the surface motions, is shown. The maximum acceleration turned out to be
0.13 g, so some amplifaction occurs. The next step would be to calculate the stress history at the surface and to represent this irregular history by an equivalent number of uniform cycles of loading
at a mean stress level [6]. Then this equivalent number has to be
compared with the results of the cyclic tests at the relevant stress
level, after which it can be judged whether this particular earthquake leads to dangerous circumstances.

In figure 11. the results of the calculations are shown in case
the sedimentary layers consist of a uniform sand with constant properties. Now the maximum acceleration at the surface were calculated at 0.18 g , so in this case a greater amplification has been
found. The shear-stress level, somewhat lower than in the foregoing
case due to the higher volume-weight of the sand, decreases more
slowly with depth than in the clay layers.

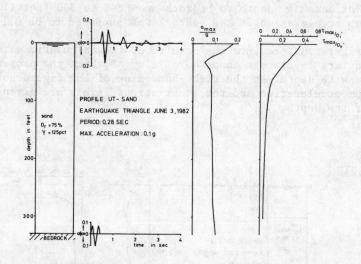

Fig. 11.   Response of sand profile.

The next calculations were made to investigate the influence of
small disturbances in the soil profile. Figure 12. presents the
results of the calculations in case at a depth of 49.5 feet a 5.5
feet thick sandlayer is present, while the remaining layers consist
of clay. The now obtained results practically coincide with those
calculated for the clay profile and shown in figure 10. However, it
is not permitted to conclude that a thin sand layer among clay layers
never influences the propagation of waves.

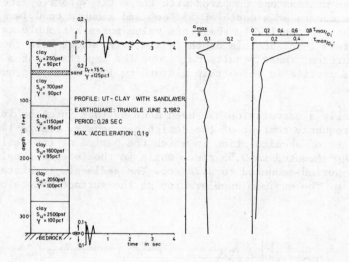

Fig. 12. Influence of sand layer to response of clay profile.

In figure 13. the results are presented of calculations in case at a depth of 49.5 feet a 5.5-feet thick clay layer is found among uniform sand with constant properties. Now the maximum acceleration at the surface only amounts to 0.054 g, so a considerable reduction has been calculated. Consequently the shear-stress level is much lower than in the foregoing cases. The acceleration at the surface

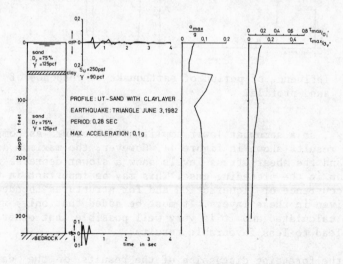

Fig. 13. Influence of clay layer to response of sand profile.

is not the maximum one compared with the entire profile. The maximum value occurs at a depth of 55 feet and amounts to 0.14 g.
However, it is not likely that this value poses a trouble as the
shear-stress level at this depth is low, viz. about half the value
at the surface. These results may prove the importance of a reliable soil profile in which thin disturbing layers are not over-
looked.

Finally a calculation has been made to look into the influence
of the frequency content of the design earthquake. Figure 14. shows
the results of a calculation in which the period of the design
earthquake amounted to 0.55 sec., while in the foregoing calculations the period amounted to 0.28 sec. The sediments consisted of
sand again. The maximum acceleration at the surface was calculated

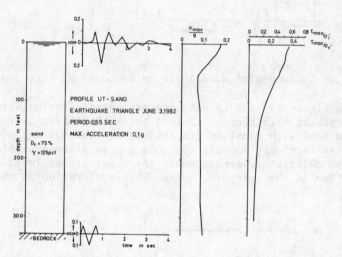

Fig. 14.   Influence of period of earthquake to response of
           sand profile.

at 0.17 g , so a somewhat lower amplification occurs as compared
with the results shown in figure 11. However, the maximum accelerations and the shear-stress levels show a slower decrease with
depth than in the preceding case. This may be important in relation
to the occurrence of liquefaction and for structures founded on
piles driven in these layers. It must be added that only one case
has been calculated and it is very well possible that other circumstances lead to less favourable results.

In the foregoing discussion of the results of the calculations
the maximum acceleration at the surface is used for comparing the
different cases only. It may be observed that looking at the maximum
acceleration is not sufficient to judge the earthquake risk at a
given site.

As already mentioned before the complete stress history at the site caused by the design earthquake must be taken into account.

It also may be observed that in the calculations, dealt with in the foregoing, only one accelerogram has been introduced. As it followed that the frequency content of the accelerogram has an influence, it is important to make a number of calculations with different accelerograms, to be sure that the right frequency content has been taken into account.

5. CLOSING REMARKS

Some information has been given about the strength and deformation characteristics of clay and of sand. Many results are available at many institutions and it would be worthwhile to compile these results. Possibly it then becomes clear that some data are not complete or that other data are missing. Also this compilation would be valuable to do preliminary calculations. The spread in the obtained results often will be such that investigations at the site will be necessary, unless the preliminary calculations, starting from conservative assumptions, show very favourable results.

Once more it is pointed out that in relation to the design earthquake the maximum acceleration, the duration and the frequency content are very important and that a reliable soil profile is a necessity. To attain this target the co-operation of a geologist, a seismologist and a geotechnical engineer is needed.

REFERENCES

[1]   Andersen, K.H., Brown, S.F., Foss, I., Pool, J.H. and
      Rosenbrand, W.F.,
      Effect of Cyclic Loading on Clay Behaviour.
      Oslo, Norwegian Geotechnical Institute, 1976,
      Publication No. 113.

[2]   Andersen, K.H., Pool, J.F., Brown, S.F. and Rosenbrand, W.F.,
      Cyclic and Static Laboratory Tests on Drammen Clay.
      Oslo, Norwegian Geotechnical Institute, 1980,
      Publication No. 131.

[3]   Schnabel, P.B., Lysmer, J. and Seed, H.B.,
      SHAKE, a Computer Program for Earthquake Response
      Analysis of Horizontally Layered Sites.
      Berkeley, University of California, 1972,
      Report No. EERC 72-12.

[4]   Housner, G.W.,
      Strong Ground Motion,
      in: Wiegel, R.L. (ed.), Earthquake Engineering.
      Englewood Cliffs, N.J., Prentice-Hall, Inc., 1970.

[5]   Newmark, N.M. and Rosenblueth, E.,
      Fundamentals of Earthquake Engineering.
      Englewood Cliffs, N.J., Prentice-Hall, Inc., 1971.

[6]   Annaki, M. and Lee, K.L.,
      Equivalent Uniform Cycle Concept for Soil Dynamics.
      Journ. Geotechn. Engng. Div.,
      Proc. ASCE, 103(1977), GT6(June), p. 549-564.

PROBLEMS IN SOIL MECHANICS AND EARTHQUAKE ENGINEERING IN THE
NORTH SEA

Per B. Selnes

Norwegian Geotechnical Institute

ABSTRACT

Differences between offshore and onshore earthquake engineering
which must taken into account in the design, are caused by the
presence of water which changes the dynamic behaviour of structures, introduces new environmental loads and complicates soil
investigation and visual site inspection. Furthermore, the structures offshore may be much larger than most onshore facilities.
Other environmental loads such as current, sea waves and ice may
act simultaneously with an earthquake. This paper discusses problems encountered offshore compared to onshore, and outlines some
of the special considerations necessary in aseismic design of
structures in the North Sea.

COMPARISON BETWEEN OFFSHORE AND ONSHORE GEOTECHNOLOGY

The tasks of the geotechnical engineer in a seismic design of
structures are to
• evaluate effects of local geology and soil on the characteristics of earthquake ground shaking;
• evaluate effects of earthquake ground shaking on stability and
  deformations of soil deposits;
• ensure safe and economic aseismic design of soil foundations and
  soil structures;
• evaluate either dynamic damping and stiffness values of the soil
  foundation, or the earthquake motion at the base of the structure
  for use in subsequent structural design.

These tasks are the same offshore as onshore. The geotechnical

219

A. R. Ritsema and A. Gürpınar (eds.), Seismicity and Seismic Risk in the Offshore North Sea Area, 219–233.
Copyright © 1983 by D. Reidel Publishing Company.

engineer working offshore draws heavily from the knowledge and
experience obtained onshore, for instance in the design of nuclear
power plants. Offshore problems are, however, in many respects
different from those encountered onshore, and experience may not
be directly transferable. Specific examples of differences which
should be accounted for in the design include: (1) the structures
offshore are often much larger than most onshore facilities; (2)
other environmental forces may be large and may act simultaneously
with an earthquake, see Fig. 1; and (3) the presence of water
changes the dynamic behaviour of structures, changes the characteristic of the earthquake ground motion and introduces new forces.

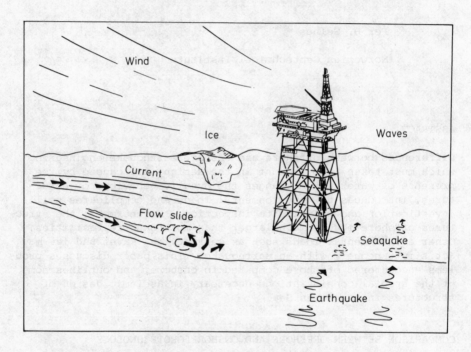

Fig. 1   Environmental forces on offshore structures

    Another major difference involves flow slides which offshore
have been found to travel several hundred kilometers over slopes
considerably below 1° (5, 13). Several large slides have been
triggered by earthquakes (18), and some older slides in the North
Sea, discovered by geophysical surveys, may also most easily be
explained by earthquake trigger mechanism (4).

    Eearthquakes furthermore generate pressure waves in the water
(seaquakes) which so far have received little attention. Seaquakes may, however, be equally disturbing or damaging offshore as

earthquakes are onshore, as for instance testified by this excerpt from a newspaper article describing the situation in Oslo harbour during an earthquake in 1904:

> "The water erupted all over, as if it has started to boil, and on board ships it felt like violent heavy seas. Simultaneously, hard blows seemed to hit against the ship's hull. Many ships already speeding on course came to a full stop, such that the crew believed that they had suddenly run ashore."

This earthquake was of magnitude 6 - 6.5, and the zone of main enerby release was 70 - 80 km away from the harbour. The intensity onshore near the harbour was about VI (MM). Several more cases of seaquakes are reported in Ref. (7).

Comparison with nuclear power plants

Many of the methods used offshore are based on the advanced procedures developed for nuclear power plants. It may, therefore, be of some interest to compare the characteristics of geotechnical problems encountered for such structures with those encountered for fixed offshore oil and gas structures:

Offshore oil gravity structure:	Nuclear power plant:
- Built where the hydrocarbons are found, very weak foundations may have to be utilized.	- Favourable site conditions selected.
- Surface or shallow foundation.	- Embedded foundation.
- Periods of interest up to 5 - 6 sec.	- Periods of interest up to 1 - 2 sec.
- Base width up to 150 m, out-of-phase motion may be important.	- Base width generally less than 60 m.
- Other large environmental loads such as sea waves, ice, current, wind; simultaneous action with earthquake possible.	- Other environmental forces negligible.
- Design may be for collapse loads; analysis of permanent and cyclic displacements desired.	- Design does not allow significant plastic yielding to occur - i.e. equivalent linear

analysis and super-
position methods are
applicable.

– Surrounding water gives added mass and damping to the structure.	– No water above the ground.
– Water transmits compressional wave causing seaquakes.	– No water above the ground.
– The presence of water complicates soil investigation and visual site inspection.	– No water above the ground.
– Relatively high degree of sample disturbance, in situ testing limited.	– High quality sampling and in situ tests possible.
– High erosion, high amount of material transport.	– No erosion.

Vertical accelerations

Vertical accelerations are more important offshore than onshore.
Static design onshore corresponds to gravity loading while offshore structures are designed for submerged weight only. The
vertical earthquake forces on the other hand, are proportional to
the mass of the structure onshore and to the mass of the structure
plus added mass from water onshore.

Figure 2 shows vertical loads on a mooring bouy. The boyancy
of the structure is balanced with the loads in such a way that the
vertical forces on the universal joint are minimized. However,
even relatively small accelerations from earthquakes will give
fairly large forces on the universal joint.

Offshore structures are, furthermore, often equipped with
large cantilever beams which are vulnerable to vertical accelerations.

Offshore soils

Soil investigations offshore are, in general, carried out to a
much smaller extent than for equally important structures onshore.
The sample quality is furthermore rarely as good as onshore due to
high water pressure (high total in situ pressure), insufficient
heave compensation during sampling, and unsophisticated sampling
techniques. Accordingly, methods to correct for sample disturbance become very important offshore. Lee (12) compares offshore

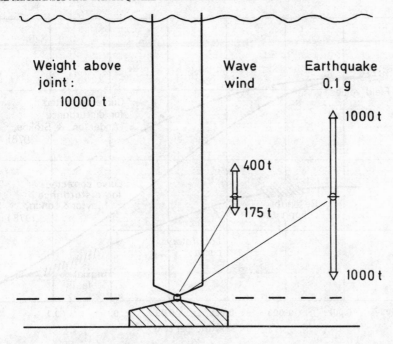

Fig. 2   Vertical forces on a mooring bouy

in situ strength results with results from tests on samples obtained with different types of corers. The strength obtained on
samples from the best types of corers was twice as high as those
obtained with regular samplers. A comparison between shear
modulus values obtained in the field and in the laboratory is
shown on Fig. 3.

    While sample disturbance may result in conservative design
for foundation stability, the design of the superstructure may be
on the unsafe side since a stiffer foundation in general yields
higher stresses in the structure for earthquake loading.

    The soils encountered offshore vary greatly and include very
soft and loose materials. Both deposition and erosion occur at a
much more rapid rate than onshore, and mobile deposits may represent a problem. Rapid deposition may also lead to the presence of
underconsolidated materials.

    Pockmarks and gasified sediments have often been found in
connection with offshore oil fields. Pockmarks in the North Sea
are found up to 300 m wide and 10 - 15 m deep. The most likely
explanation is the release of gas from deeper sediments or gas
reservoirs (8).

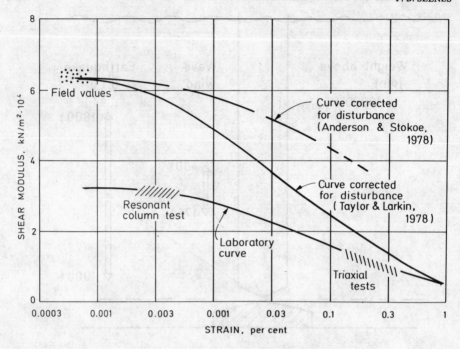

Fig. 3   Comparison between field and laboratory shear modulus
          values for a slightly overconsolidated clay

The presence of gas in the sediments tends to (1) increase
the sample disturbance, (2) interfere with in situ measurements,
(3) influence the pore pressure in the soil, and (4) change the
behaviour under loading (6, 14).

OTHER ENVIRONMENTAL LOADS

Sea waves will in general cause the higher loads on the structure
for regions with relatively low seismic activity such as the North
Sea. Earthquake forces may, however, still be important and may
govern the design for certain types of structures and structural
components (16). Table 1 shows displacement values computed for
sea waves and earthquakes for approximately the same, low level of
probability. Note the different factors between the 10^{-2} and the
10^{-4} level loading for the two load cases - i.e. earthquakes become relatively more important for low probabilities.

The maximum forces on a typical North Sea gravity structure
from the design wave (31 m) are shown on Fig. 4. The period of
motion is generally within 5 - 20 sec. for larger waves, and the
energy is concentrated in a very narrow band compared to earth-

Load	Cyclic displacement		Cyclic strain
	Horizontal	Vertical	
Earthquake ~ 4 • 100 year quake	± 5.5 cm	± 6.5 cm	± 0.5 %
Sea wave ~ 1.3 • 100 year wave	± 3.5 cm	± 4.0 cm	± 0.25%

Table 1   Comparison between computed displacements and strain for
          earthquake and sea wave loading for approximately 10^{-4}
          level probability. Vertical displacements are for the
          platform edge, strains are given for an element immedi-
          ately beneath the center of the platform base. Displace-
          ments are relative to the soil base at 80 m depth (17).

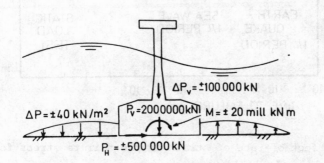

Fig. 4   Forces on a gravity structure from the 100-year design
         sea wave. The foundation area is 6300 m^2 (15).

quake loading. Wind forces are relatively small and may be only
10 - 20% of the forces from the sea wave. Sea current forces are
also usually relatively minor.

     Important differences between sea wave and earthquake loading
are rate of loading (Fig. 5) and number of load cycles. Earthquakes in the North Sea may be expected to have less than 10 significant cycles while sea waves have several hundreds.

ANALYSES

The selection of method of analysis must be based on the level of
strain developed in the soil during the earthquake. While linear
elastic analyses of soil behaviour may be acceptable provided the
level of strain is low enough, nonlinear effects become very important at higher strain and must be taken into account.

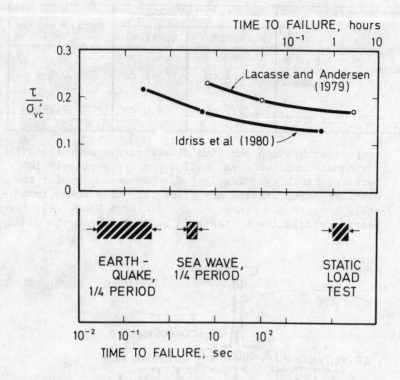

Fig. 5   Effect of rate of strain on the failure stress for clay

Nonlinear effects and the required increase in degree of
sophistication in the analytical modelling of the soil with increase in strain level are illustrated in Fig. 6.

The level of shaking where nonlinear effects must be taken
into account depends on the type of soil. Figure 7 shows change
in stiffness from the 1st to the 10th cycle in stress controlled
simple shear tests. Procedures based on constant stiffness for
the whole shaking period are obviously not valid for cyclic shear
strains above 0.1 percent strain for loose sand, 0.3 - 0.5 percent
for dense sand and normally consolidated clay, and 1 - 2 percent
for overconsolidated clay.

Furthermore, procedures based on linear elasticity cannot
give the magnitude of permanent (irrecoverable) deformations of
the foundation during the earthquake.

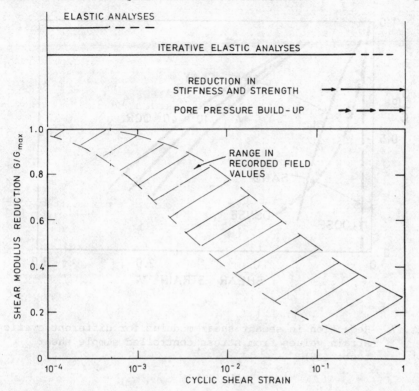

Fig. 6   Nonlinear effects and methods of analyses for various
          strain levels. Recorded field values are from Arango
          et al. (3)

Displacement analyses

Analyses of two platform foundations, both with low estimates of
soil stiffness and strength, carried out for upper bounds of earthquake loading in the North Sea (0.25 g horizontal and 0.15 g vertical accelerations), gave following results:

• Cyclic displacements were between 2.5 and 6.5 cm.

• Permanent displacements were generally less than 4 cm, except
  for 20 cm displacement in vertical direction in one of the cases.

• Cyclic and permanent strains beneath the platform were less than
  0.5 and 1 percent, respectively.

• Maximum horizontal accelerations were reduced from 0.25 g in the
  soil base (control motion) to 0.035 - 0.05 g in the concrete base
  of the platform.

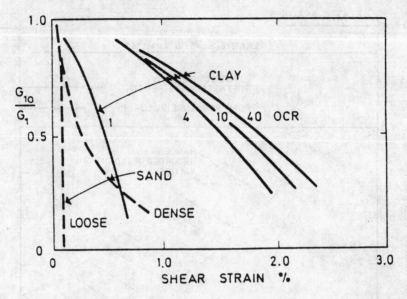

Fig. 7   Reduction in secant shear modulus for different cyclic
         strain values from stress controlled simple shear
         tests (18)

All displacements are relative between the concrete base of
the platform and the soil base at 80 m depth. Examples of computed displacement-time histories are shown on Fig. 8 (17).

Other nonlinear analyses with various degrees of simplifying
assumptions have given similarly small and quite tolerable foundation movements even for earthquake loads pushing the soil stresses
well into the failure range (10, 21). If this is the true behaviour of offshore structures, the necessary ductility to take large
earthquake forces can be provided by the foundation, thus reducing
the forces on the structures. Such a "soft first floor" design
may prove to be more easily adapted offshore than onshore. There
is, however, still a long way to go before nonlinear analytical
tools are developed and verified to an extent that allow such
design concepts to be fully utilized.

PROBLEMS IN ASEISMIC DESIGN FOR THE NORTH SEA

Most of the knowledge about strong ground motions from earthquakes
has been obtained from regions in the world where seismic, tectonic, and geologic conditions are quite different from those in the
North Sea. Differences which may have pronounced effects on the
characteristics of strong ground shaking in the North Sea are:

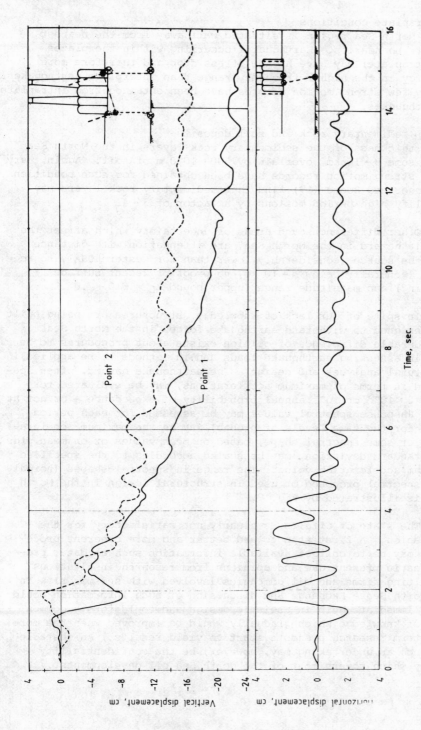

Fig. 8   Vertical and horizontal displacements of platform concrete base relative to rock base for normally consolidated clay foundation (17)

- Intraplate conditions
  The North Sea region is situated well away from the nearest
  plate boundaries. Earthquakes occurring within the plates
  (intraplate) may have higher stress drop and therefore more
  energy in the higher frequency range than interplate earthquakes.
  Very few strong motion records have been obtained from intraplate
  earthquakes.

- Deep sedimentary rock and soil deposits
  The thickness of the sedimentary rock layers in the North Sea
  are some 8 - 10 km, overlain by 500 - 1000 m of soil. Again, very
  few strong motion records have been obtained for such conditions.
  Swanger and Boore (19) find that sedimentary rock layers may
  amplify high period motions by a factor of 3 - 4.

Other differences and areas of uncertainty which affect the
seismic hazard in the North Sea, are attenuation with distance
from the source (considerably lower than for western USA), source
depth (generally below 5 - 10 km, no proof of recent surface
faulting) and magnitude range (upper bound $M_L = 6.0 - 7.0$).

In spite of our lack of knowledge, structures are being built
and designed to withstand earthquake forces in the North Sea.
Considerable difference of opinion exists about procedures to be
used in assessing earthquake loads and in methods to be applied in
structural analyses and design. The earthquake hazard, often specified in terms of maximum accelerations, may be evaluated for
10^{-2}, $2 \cdot 10^{-3}$ or 10^{-4} annual probability of exceedance - or nor at
all. Response spectral values may be assessed for each period
range for the same levels of probability as the maximum acceleration; or some spectral shape, based on mean values or on mean plus
one standard deviation, may be scaled according to the specified
maximum acceleration value. The range in spectral shapes (normalized spectra) proposed or used in structural design in the North
Sea, is illustrated by Fig. 9.

The state of affairs is clearly not satisfactory for the
profession. A first step toward better and more coherent procedures may be to collect available information such as data, procedures in present use and opinions from research institutions,
consulting firms and oil companies involved with seismic risk in
the North Sea. Exchange and discussion of such information would
be of immediate help in aseismic design and could form a common
base of knowledge which gradually would be improved when the more
long term research projects start to yield results. An obstacle
for such an undertaking may, however, be the confidentiallity
policy which covers most of the North Sea oil developments.

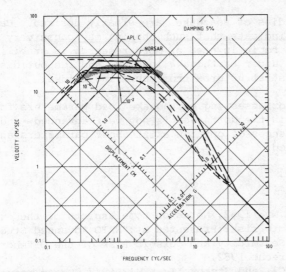

Fig. 9   Normalized response spectra proposed for deep competent
         soil deposits in the North Sea

SUMMARY AND CONCLUSION

The presence of water in the offshore environment has considerable
implications for the earthquake analyses. The water complicates
the site investigations, increases sample disturbance, changes the
characteristics of the sea floor soils, changes the earthquake
ground motion, transmits compressional waves (seaquakes), changes
the dynamic characteristics of structures and introduces additional large environmental forces such as storm waves and ice.

     Offshore structures are in many respects quite different from
most onshore structures. Large foundation dimensions make out-of-
phase motions from horizontally travelling waves more important.
Large structures with additional mass from the surrounding water
have high foundamental periods, and duration and higher period
components of the earthquake ground motion become more important.
Reduced static loads in vertical direction due to the boyancy of
submerged structures increase the importance of the vertical
earthquake component.

     Analytical procedures developed for onshore problems are ex-
tensively used offshore. However, soil-structure-interaction may
be more important for offshore structures due to the large dimensions and large masses. Nonlinear analyses may be more desirable
offshore since studies seem to indicate that even very serious
earthquake loading can be sustained with limited movements of the
structure, and, furthermore, that yielding in the soil foundation
may reduce the forces on the structure.

Further studies of seaquakes are clearly desired. Case studies seem to indicate that such shaking may cause very serious damage. Another topic in need of more research is stability of submarine slopes under earthquake loading, and run out distance and forces against structures from submarine slides.

The basis for assessing earthquake load parameters for aseismic design in the North Sea is not well advanced, and considerable differences of opinion exist about parameters and procedures to be used.

References:
1. Ambraseys, N.N.: 1982, Notes on some aspects of the evaluation of seismic risk. Presented at: "NATO Advanced Study Workshop on the Seismicity and Seismic Risk of the Offshore North Sea Area."Utrecht 1982.
2. Anderson, D.G., and Stokoe II, K.H.: 1978, Shear modulus; a time-dependent soil property. "American Society for Testing and Materials. Dynamic geotechnical testing; a symposium. Denver, Colorado 1977. ASTM Special technical publication, 654," pp. 66-90.
3. Arango, I., Moriwaki, Y., and Brown, F.: 1978, In-situ and laboratory shear velocity and modulus. "American Society of Civil Engineers. Specialty Conference on Earthquake Engineering and Soil Dynamics."Pasadena, California 1978. Proceedings, Vol. 1, pp. 198-212.
4. Bugge, T., Lien, R.L., and Rokoengen, K.: 1978, Kartlegging av løsmassene på kontinentalsokkelen utenfor Møre og Trøndelag; seismisk profilering."Institutt for kontinentalsokkelundersøkelser,"Theim. Publikasjon, 99. 55p.
5. Embley, R.W., and Jacobi, R.D.: 1977, Distribution and morphology of large submarine sediment slides and slumps on Atlantic continental margins. Marine Geotechnology, Vol. 2, Marine Slope Stability Volume, pp. 205-228.
6. Esrig, M.I., and Kirby, R.C.: 1977, Implications of gas content for predicting the stability of submarine slopes. Marine Geotechnology, Vol. 2, pp. 81-100.
7. Hove, K., Selnes, P.B., and Bungum, H.: 1982, Seaquakes; a potential threat to offshore structures. Presented at: "International Conference on the Behaviour of Offshore Structures, 3. BOSS'82."MIT, Cambridge, Mass. 1982.
8. Hovland, M.: 1981, Characteristics of pockmarks in the Norwegian Trench. Marine Geology, Vol. 39, No. 1/2, pp. 103-117.
9. Idriss, I.M., Dobry, R., and Power, M.S.: 1975, Soil response considerations in seismic design of offshore platforms. "Offshore Technology Conference, 7."Houston 1975. Proceedings, Vol. 3, pp. 192-205.

10. Kausel, E.A., Lucks, A.S., Edgers, L., Swiger, W.F., and Christian, J.T.: 1979, Seismically induced sliding of massive structures. American Society of Civil Engineers. Proceedings, Vol. 105, No. GT 12, pp. 1471-1488.

11. Lacasse, S., and Andersen, K.H.: 1979, Effect of load duration on undrained behaviour of clay and sand; summary of literature and laboratory studies and recommendations for modified laboratory procedures. "Norwegian Geotechnical Institute. Internal report, 40007-4." 20p.

12. Lee, H.J.: 1979, Offshore soil sampling and geotechnical parameter determination. "Offshore Technology Conference, 11." Houston 1979. Proceedings, Vol. 3, pp. 1449-1458.

13. Moore, D.G.: 1978, Submarine slides. "Rockslides and avalanches. Vol. 1: Natural phenomena. Ed. by B. Voigt."Amsterdam, Elsevier. Pp. 563-604.

14. Sangrey, D.A.: 1977, Marine geotechnology - State of the art. Marine Geotechnology, Vol. 2, pp. 45-80.

15. Schjetne, K., Andersen, K.H., Lauritzsen, R., and Hansteen, O.E.: 1979, Foundation engineering for offshore gravity structures. Marine Geotechnology, Vol. 3, 1979, No. 4, pp. 369-421."Norwegian Geotechnical Institute. Publication, 129."

16. Selnes, P.B., Ringdal, F., Jensen, D., and Hove, K.: 1980, Earthquake risk on the Norwegian Continental Shelf. Norwegian Maritime Research, Vol. 8, No. 1, pp. 13-25.

17. Selnes, P.B., Rognlien, B., and Olsen, T.S.: 1982, Cyclic and irrecoverable displacements of offshore platform foundations during earthquakes. "Norwegian Geotechnical Institute. Report, 40009-5." Var.pag.

18. Selnes, P.B.: 1981, Geotechnical problems in offshore earthquake engineering. "Norwegian Geotechnical Institute. Internal report, 40009-6." 60+12p.

19. Swanger, H.J., and Boore, D.M.: 1978, Inportance of surface waves in strong ground motion in the period range of 1 to 10 seconds. "International conference on Microzonation for Safer Construction-Research and Applicaiton, 2."San Fransisco, California 1978. Proceedings, Vol. 3, pp. 1447-1457.

20. Taylor, P.W., and Larkin, T.J.: 1978, Seismic site response of nonlinear soil media. American Society of Civil Engineers. Proceedings, Vol. 104, No. GT 3, pp. 369-383.

21. Watt, B.J., Boaz, I.B., Ruhl, J.A. Dowrick, D.J., and Ghose, A.: 1978, Earthquake survivability of concrete platforms. " Offshore Technology Conference, 10."Houston 1978. Proceedings, Vol. 2, pp. 957-973.

SEISMIC DATA REQUIREMENTS FOR THE DESIGN OF FIXED PLATFORMS IN THE OFFSHORE NORTH SEA AREA

OVE T. GUDMESTAD,

LEAD STRUCTURAL ENIGNEER, A/S NORSKE SHELL*

*   Employee of Statoil, Stavanger seconded to A/S Norske Shell's Engineering Department.

ABSTRACT

The necessary data and calculations for the earthquake design of fixed platforms in the offshore North Sea Area are being reviewed. The relevant Norwegian Codes are being discussed.

SUMMARY

-   For the general earthquake design of fixed platforms, the following data and calculations are necessary:

-   Peak ground acceleration or velocity at bedrock

-   Full desciption of the earthquake loading at bedrock at the specific site:
    * relevant response spectrum
           or
    * relevant time histories

-   Free field earthquake properties, i.e. modification of the earthquake due to the local soil conditions.

-   Soil- structure interaction modelling

-   Design checks according to the relevant code(s)
    * linear checks for low intensity earthquakes
    * ductility checks or nonlinear checks for higher intensity earthquake levels

235

A. R. Ritsema and A. Gürpınar (eds.), Seismicity and Seismic Risk in the Offshore North Sea Area, 235–248.
Copyright © 1983 by D. Reidel Publishing Company.

- Relevant Norwegian codes are reviewed

- Suggested design checks are discussed

1. <u>INTRODUCTION</u>

A/S Norske Shell has partnership interests in several licences in the Norwegian Sector and acts as operator for some of these licences, amongst these is licence 054, block 31/2.

Block 31/2 is located approximately 80km north west of Bergen in the Norwegian Trench (Fig. 1), and contains a large accumulation of sweet lean gas which is underlain by a thin oil layer. The development of these reserves will have to cope with significant technical challenges, particularly the water depth (320 m to 350 m), the soft seabed conditions and the shallowness of the reservoire which extends over a large area. Various field development plans are currently being formulated and use of fixed platforms are being studied in combination with or as alternative to floating production facilities. Amongst the fixed platforms under study are Norwegian Contractor's Condeep T300, Heerema's steel Tripod Tower Platform (TTP) and a Piled Tower Platform (jacket), conf. Fig. 2.

2. <u>SEISMIC DESIGN CONSIDERATIONS FOR DEEP WATER FIXED PLATFORMS</u>

Platforms for deep water with a first natural period up to 5-6 seconds are more sensitive to the low frequency range of the earthquake exitation than platforms for more conventional waterdepths having first natural periods typically in the 2-3 second range.

The peak value of the spectral acceleration curve is found in the 0.15 to 0.70 sec range (1). For deep water platforms the higher natural periods are located in this range. It can thus be expected that local overstressing due to excitation of local modes becomes more severe than for platforms designed for shallower water. The combination of the responses has to include many modes for the deep water platforms. The uncertainties in the calulation of the higher modes and the applicable modal superposition rules require further studies to obtain reliable results. Time history techniques may have to be used instead of spectral techniques (5).

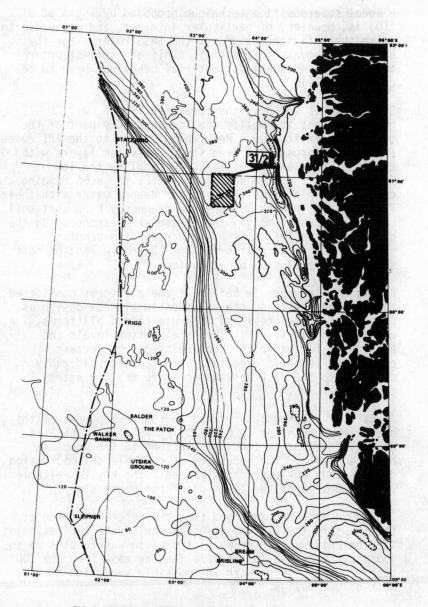

FIG. 1 NORTHERN NORTH SEA
LOCATION MAP, NORWEGIAN SECTOR

A modal superposition technique proposed by Vugts et al (10) is, however, of particular interest, and may prove to solve the deficiencies of eg. the SRSS rules. For the higher earthquake intensities, nonlinear soil-structure interaction and nonlinear material behaviour have to be considered (4).

Soft soil, as found in the Norwegian Trench, will generally tend to amplify the low frequency part of the earthquake excitation. Reference is made to the API rules (1) which propose that deep strong alluvium layers with thicknesses in excess of about 60m amplify the low frequency part of the spectrum by 2.25 for a 5% damping case . For platforms designed for deeper water with their first natural periods in the low frequency band soft soil may considerably amplify the earthquake response. On the other hand the soft soil will increase the damping experienced by the platforms, and the net effect of soft soil still remains to be evaluated.

Due to the increase in height of the platforms considered, their slenderness and the unconventional deck design as proposed for the steel and concrete tripod platforms, i.e. use of long cantilevers, the acceleration levels and associated forces in the deck related to a certain earthquake level is expected to increase considerably compared to the deck designs for more shallow water platforms.

As part of the studies related to the development of the hydrocarbon reserves in block 31/2 A/S Norske Shell is heavily involved in finding solutions to the problems outlined above. The necessity of reliable seismic design data is thus obvious, and represents the key to reliable seismic design of the fixed platforms under study.

It should also be mentioned that the general opinion of the industry is that the expected earthquake excitation of the conventional North Sea platforms in waterdepths up to 150 m represents a less severe loading case than the 100 year storm condition with the possible exception of the deck-column design.

3.   NORWEGIAN CODES

The Norwegian Petroleum Directorate's rules (7) require checks of platform behaviour for an accidental earthquake level with a probabiliy of exceedance of 10^{-4} per year.

"The Norwegian Petroleum Directorate do not require a detailed analysis documentation for specified design accidental events and effects. An engineering approach based on evaluation of actual damage potential, experience, possible historical data, and reliability data for the systems will normally be sufficient. However, if the Norwegian Petroleum Directorate consider the specified accidental effects to be unreasonable, further clarification and justification of the values may be required in the detailed design phase" (7), Article 4.2.5.

The Norwegian research institute Norsar specializing in earthquake prediction is currently for the Norwegian "Safety Offshore" program preparing zoning maps offshore Norway showing expected peak ground accelerations in bedrock at a probability of exceedance equal to 10^{-4} and 10^{-2} per annum (9). Preliminary results are given in Fig. 3. The uncertainties in the presented figures are appreciable, and the given maps should only be used for preliminary assessments of earthquake risks in the Norwegian Sector. The earthquake having a 10^{-2} probability of exceedance per annum is suggested to be treated as an environmental load according to the NPD rules (6).

The design principles of the Norwegian Petroleum Directorate's rules request that

"Structures and structural elements shall be designed to

-    Sustain all loads and deformations with an adequate degree of safety against failure

-    Maintain, in case of accidental loads causing local failure, adequate safety against progressive collapse"

Four categories of "limit states" are defined:

-    "The ultimate limit states, related to the risk of failure or large inelastic displacements or strains of a failure character

-    The serviceability limit states, related to criteria governing normal use or durability

-    The fatigue limit states, related to the risk of failure due to the effects of repeated loading

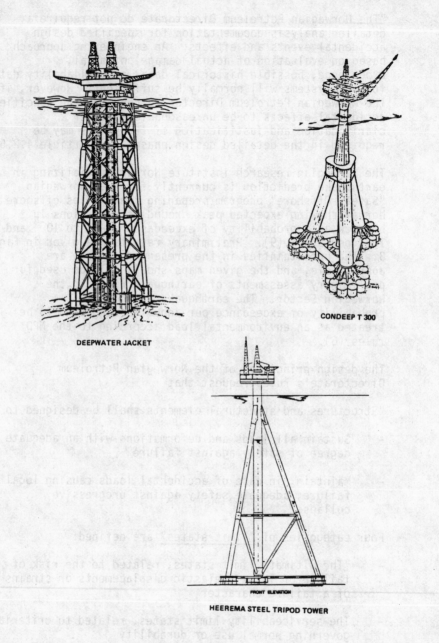

DEEPWATER JACKET

CONDEEP T 300

HEEREMA STEEL TRIPOD TOWER

Possible fixed platforms for development of block 31/2 in the
Norwegian Sector of the North Sea.

- The limit states of progressive collapse, related to the risk of failure of the structure under the assumption that certain parts of the structure have ceased to perform their load-carrying functions".

(6), Chapter 2

The objectives of the design analysis are to keep the probability of a limit state being reached below a certain value.

The regulations are based on verification of adequate safety by ascertaining that the <u>design loading effects</u> satisfy given <u>resistance criteria.</u> Design loading effects are based on design loadings, found by multiplying characteristic loads by <u>load coefficients, γ_f</u> . Similarly, design resistances are found by dividing characteristic resistances by <u>material coefficients, γm.</u>

The characteristic value of an <u>environmental load</u> is defined as the most probable value with annual probability of exceedancy of 10^{-2}. The ultimate limit state and the serviceability limit state are based upon the same characteristic load, if no other procedure is described in the rules.

The load coefficients, γf, are defined as follows:

- The ultimate limit state shall be checked for two load combinations, ordinary and extreme, by using load coefficients according to Table 3.1.

Load combination	P	L	D	E	*
Ordinary	1.3	1.3	1.0	0.7	
Extreme	1.0	1.0	1.0	1.3	

Table 3.1 : Load coefficients for the ultimate limit state.

* P - permanent loads
  L - live loads
  D - deformation loads
  E - environmental loads

The loads shall be combined in the most unfavourable manner, provided that the combination is possible for physical reasons and allowed by loading specification.

-   For the serviceability limit states, the design loads are normally equal to the characteristic loads, i.e. for all loads $\gamma f = 1.0$

-   For the fatigue limit states loading effects will occur with variable amplitudes in random order. These loading effects may be calculated by using a load coefficient $\gamma f = 1.0$.

-   The limits state of progressive collapse shall be checked for accidental loads. A local failure is acceptable provided that the remaining structure satisfies the requirements in the ultimate limit state for the load combinations given in Table 3.2. The load combination "Accidental load included" shall only be checked if the accidental load physically can occur after the local failure. Only one accidental load is included in the combination.

Load combinations	P	L	D	E	A **
Accidental load included	1.0	1.0	0	0	1.0
Accidental load excluded	1.0	1.0	0	1.0	0

Table 3.2:   Load coefficients for progressive collapse.

** A   -   accidental load

4.   SEISMIC DATA REQUIREMENTS

General seismic risk maps as proposed by API for United States Coastal waters (1) and Norsar (9) for the Norwegian Sector (Fig. 3) are useful as general guidelines. However, when designing platforms for specific sites, site specific studies are recommended.

The local tectonics of the site and adjacent area must be taken into account for more accurate definition of the earthquake parameters to be used for further analysis of the platform behaviour subject to the earthquake exitation.

A significant amount of work has been done over the past decade to develop probabilistic models for evaluation of seismic exposure. A source seismicity model that characterizes seismic sources including location, geometry (point sources or faults with physical extension) and earthquake recurrence taking into account historic data is the basis for the model. The definition of an attenuation model that provides estimates of ground motion values at various distances from the sources has to be established. A procedure that combines the contribution of all sources to give the seismic exposure at a specific site finalizes the model. The peak ground acceleration or velocity levels at bedrock for the specific site as a function of the recurrence period should be the outcome of the model. For engineering purposes the most probable values with certain recurrence periods are required. Standard deviations for the values should be given to facilitate the evaluation of the degree of conservatism involved in the proposed criteria. Through development of a refined model a better understanding of North Sea tectonics, more knowledge about the geology of the areas and further collection of earthquake data, less conservatism can be justified with possible savings in the structural design. Given the peak ground acceleration level, i.e. the acceleration response at very high frequencies, the full description of the earthquake loading at the baserock is given through a response spectrum or appropriate time histories scaled to the zero period acceleration level. The API rules (1) suggest a response spectrum for rock and recommend that in the case of the time history approach at least three time histories should be considered in the design.

For the North Sea Area recordings of strong motion earthquakes are scarce. Either the suggested spectra or the recommended time histories must be taken from areas with similar tectonic structure. The uncertanties in the establishment of the spectra are large, and it may therefore be more appropriate to select earthquake time recordings relevant for intraplate tectonics and from areas with similar base rock properties.

These recordings should be from earthquakes having epicenters as expected for North Sea Area earthquakes. Thus, the frequency content of the selected time histories will be relevant for the North Sea Area.

The energy stress-relieving mechanisms for the different earthquake levels have also to be considered in the choice of spectrum or time histories. The higher acceleration level earthquakes are often considered to have more energy for the lower frequencies. (8).

Due to the potential sensitivity of the platform response to variations in the input motion, the use of the spectrum in the design should cover the acceleration levels obtained by use of the selected time histories. The spectrum of a specific time history should thus have lower values than the design spectrum which will represent the average over many time histories plus a standard deviation.

It is hoped that this workshop will shed further light into the use of spectrum versus time histories in the design and that spectra and earthquake recordings relevant for the North Sea Area can be recommended.

The API rules (1) recommend different spectra for different types of top soil. However, a more consistent approach will be to calculate the influence of the local geological and geotechnical conditions on the spectra or the time histories defined at the base rock. The possible amplification of the low frequency content of the earthquake by soft top soils might be particular critical for deep water platforms as outlined in Chapter 2. Different free field models might be applied for shear waves (horizontal motion) and for push waves (vertical motion). The obtained free field earthquake loading will be modified by the presence of a platform. A soil-structure interaction model has to be established to find the soil properties at the base of the platform /at the top of the piles.

The soil model may be supplied with "transmitting" boundaries along the lateral edges and "viscous" boundaries on the planar surfaces, to simulate the absorption of wave energy emanating from the model as e.g. done in the Flush computer code (3).

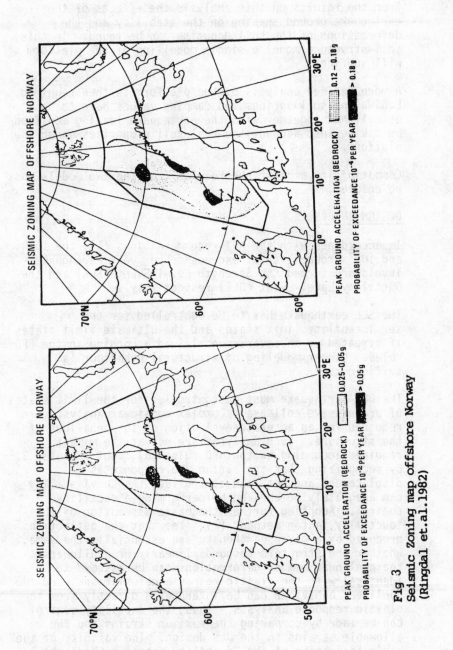

Fig 3
Seismic zoning map offshore Norway
(Ringdal et.al.1982)

From the results on this analysis the effects of the
earthquake ground shaking on the stability and the
deformations of the soil deposits can be found. In this
soil-structure model a simple modelling of the platform
will suffice.

A more refined analysis of the platform is then required.
Equivalent soil springs and damping values have to be
established together with the earthquake loading modified
by taking into account the real soil properties under the
platform.

Compatibility amongst the results from the two models must
be ensured.

5.   DESIGN CHECKS

In accordance with the API recommendations (1), the 10^{-2}
and 10^{-4} probability of exceedance per year earthquake
levels are termed the Strength Level Event (SLE) and the
Ductility Level Event (DLE) respectively.

The SLE earthquake has to be controlled for the
serviceability limit states and the ultimate limit states
if streated as an environmental load according to the NPD
rules. Linear modeling of structural behaviour is
sufficient.

The DLE earthquake must be controlled for the limit states
of progressive collapse. Complex nonlinear analyses are
required for an accurate evaluation of the behaviour of
the structure. Such analyses are not at the moment
required according to the NPD rules (6), conf. Chapter 3.
By recognizing that the earthquake response is
displacement and not load controlled, the survivability
can adequately be checked by performing a ductility
control. Adopting further the basic assumption of the
"ductility factor method" (2), i.e. that the deflections
produced by a given earthquake are essentially the same,
whether the structure responds linearly or nonlinearly,
the nonlinear member deformations can be assumed to be
identical with the elastic response deformations, and the
nonlinear behaviour can be interpreted directly from the
elastic response analysis. Thus, the ductility control
can be made by comparing the maximum strains and the
allowable strains in the ULS design. The validity of the
basic assumption of the "ductility factor method" should,
however, be further analysed. For steel tubular members
nonlinear methods are available (4).

The validity of simplified methods assessing energy absorption capacities should also be evaluated for possible replacement of complex nonlinear analyses, which as yet have not been generally developed.

6. ACKNOWLEDGEMENTS

The author expresses his thanks to A/S Norske Shell for permission to publish the paper. However, the author wishes to emphasis that the statements given represent the author's personal views, and they do not necessarily represent the official opinion of A/S Norske Shell.

References

1. API, 1981, "API Recommended Practice for Planning, Designing and Constructing Fixed Offshore Platforms". API RP2A, Twelfth Edition, January 1981.

2. Clough, R.W. and Penzien, J.: Dynamics of Structures". Mc Graw-Hill, 1975.

3. Lysmer, J., Udaka, T., Tsai, C.F. and Seed, H.B.: "FLUSH"; a computer program for approximately 3-D analysis of soil-structure interaction problems". Univ. of California, Berkely College of Engineering, Earthquake Engineering Research Center. Report EERC 75-30, 1975.

4. Marshall, P.W.: "Design Considerations for Offshore Structures having Nonlinear Response to Earthquakes". Paper presented at ASCE Convention, Chicago, October 1979

5. Nair, D., Valdivcesco, J.B. and Johnson, C.M, "Comparison of Spectrum and Time History Techniques in Seismic Design of Platforms". Journal of Petroleum Technology, March 1981.

6. Norwegian Petroleum Directorate, 1977, "Regulations for the structural design of fixed structures on the Norwegian Continental Shelf". (Unofficial translation). NPD, Stavanger, 1977.

7. Norwegian Petroleum Directorate, 1981, "Guidelines for Safety Evaluation of Platform Conceptual Design." (Unofficial translation). NPD, Stavanger, 1981.

8. Ove Arup & Partners, 1980, "Earthquake Effects on Platforms and Pipelines in the UK Offshore Area". Report for UK Department of Energy, London, May 1980.

9. Ringdal, F., Husebye, E.S., Bungum, H., Mykkeltveit, S. and Sandvin, O.A., 1982, "Earthquake hazard offshore Norway". A study for the NTNF "Safety offshore committee". NTNF/NORSAR, 1982.

10. Vugts, J.H., Hines, I.M., Nataraja, R., Schumm, W., 1979, "Modal superposition in Direct Solution Techniques in the Dynamic Analysis of Offshore Structures", 2nd International Conference on BOSS, Paper 49, London, 1979.

EVALUATION OF STABILITY OF NORTH SEA SITES

Derek W. F. Senner

McClelland Engineers, London

ABSTRACT

This contribution is a brief outline of site stability evaluation techniques with reference to typical North Sea soil conditions. Elements of site evaluation are outlined, engineering methods described for preliminary assessment of soil liquefaction potential and slope stability during earthquake loading, and results of analyses from North Sea sites discussed. Generally soils on the North Sea continental shelf appear inherently stable. However, potential for slope failures during earthquakes may exist on the continental slope.

INTRODUCTION

Evaluation of offshore site stability with respect to seismic activity comprises three main elements, namely: 1) Geophysical Survey, 2) Geological Interpretation and 3) Geotechnical Engineering. The first two of these elements provide information for comparison with sites where information is available on soil performance during earthquakes. As yet there are no reports in the literature of seimic-induced seafloor instability on the North Sea shelf, but there is evidence of slope failures, possibly induced by earthquakes, on the western slope of the Norwegian Trench, and on the continental slope in the Norwegian Sea. Typical North Sea soil conditions are presented and analysis of liquefaction potential and slope stability performed that illustrate the inherent stability of North Sea soils.

A. R. Ritsema and A. Gürpınar (eds.), Seismicity and Seismic Risk in the Offshore North Sea Area, 249–259.
Copyright © 1983 by D. Reidel Publishing Company.

SITE STABILITY EVALUATION

Geophysical Survey

A very important aspect of site stability evaluation is the geophysical survey. Detailed bathymetry is measured using a precision echo sounder and side-scan sonar is used to identify seabed topographical features. In soft sediments high resolution profiling in the upper few metres is undertaken using a tuned transducer, below which information is obtained from an acoustipulse system. In compact and granular sediments a combination of Sparker and an acoustipulse system should be used. Figure 1 shows a large slump feature detected from Sparker records. Since the prime objective of the survey is to investigate regional stability, the survey should extend to an area at least 10 km square around a proposed offshore platform site.

Geological Studies

Geological studies are undertaken primarily to assess the age of soil deposits, the depositional history and subsequent loading conditions. Age of soils is particularly important when assessing liquefaction potential since most reported cases of liquefaction are generally in young deposits. Results from laboratory tests indicate that liquefaction potential reduces with increasing application time of consolidation pressures (1).

Geotechnical Studies

Geotechnical studies are undertaken to assess the magnitude of potential site stability problems identified by geophysical survey and the geological evaluation. Generally the separate assessment of seismic loading and soil resistance will lead to a conservative estimate of site stability. If geophysical and geological studies indicate a region is historically stable, then provided there have been no significant changes in the geological setting, a site within the region is likely to remain stable for free field conditions.

EVIDENCE OF SITE INSTABILITY

The two major catagories for site stability assessment relate to a) slope stability, and b) stability of level sites. There are many examples of seafloor instability reported in the literature. Generally, such failures occur in regions of rapid deposition where sediments are either loose sands or silts, or underconsolidated soft clays. Not all such instabilities are earthquake-induced and may more frequently be caused by a combination of many factors including gravity loads, wave-induced seafloor pressures, currents and rapid draw-down during sea level lowering. Several massive submarine slides triggered by earthquakes have been reported, (2), (3).

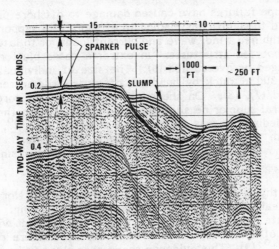

Fig. 1 10 kilojoule sparker record of a rotational slump

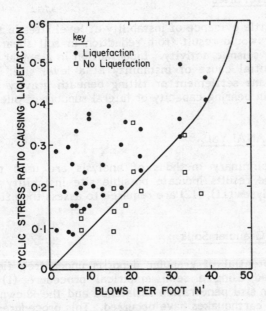

Fig. 2 Liquefaction and penetration resistance of sand (After Seed 1979)

Submarine Slope Failures

Submarine slope failures have caused damage to offshore pipelines (4), (5) and could lead to significant loading to offshore pile supported structures exceeding combined wind, wave and current loads, or to lateral movement of seabed supported structures. In extreme cases platforms have been severly damaged or overturned (6). The size of individual slides can be great; slides 300 m thick by hundreds of square kilometres in plan have been reported (7), (8). On the continental slope at Storegga, offshore Norway, several submarine slides have occurred (9). In the largest some 800 km^3 of sediments was moved on a slope of 1 to 2°, and the slide removed both Quaternary and Tertiary sediments. Analysis has apparently shown that the most likely trigger mechanism was tectonic activity.

A study of bathymetric maps on the western side of the Norwegian Trench (10) shows breaks in slope angles indicative of slumping. On the same slope, elongate scars, sub-parallel to the slope and with downslope displacement of a few metres have been observed where regional slopes are less than 1°, (9). This evidence shows that slope failures on different scales are present on the slopes of the Norwegian Trench, and may also occur on the Continental Slopes of the northern North Sea.

Instability of Level Sites

There is very little evidence of instability at level sites in the North Sea. Such instability would result from reduction in soil shear strength and stiffness during seismic activity. Liquefaction in granular deposits is an important potential cause of instability at a level site. Effects may include permanant settlement or tilting beneath gravity structures, or temporary loss in bearing capacity or lateral support of piles, pipelines or gravity bases.

ENGINEERING ANALYSIS

In practice preliminary methods of analysis are used to assess site stability. If the results indicate possible site instability then detailed engineering analyses (11), (12) are required to assess the effect of seismic activity.

Liquefaction of Granular Soils

Liquefaction potential of granular deposits under free field conditions may be assessed using a semi empirical procedure (1) based on a correlation of in situ penetration resistance and the known behaviour of locations where earthquakes have occurred. This procedures is limited to sands containing less than about 10 percent of silt sized material. Use of the procedure involves comparison of earthquake-induced seismic stress and Standard Penetration Test (SPT) blow counts, N, (13) with data presented on Figure 2, for earthquakes up to Magnitude 8.5. The indicated lower bound curve provides a conservative assessment of the data.

The in situ condition of granular deposits in the North Sea is generally measured by static cone penetrometer tests. Cone point resistance, q_c in MPa, may be converted to equivalent SPT N values, blows per foot, using an approximate correlation (14) where:

$$N = 2.5\ q_c$$

Cyclic stress ratio is computed for a given ground acceleration using the following relationship:

$$\frac{\tau_{av}}{\sigma'_v} = 0.65\ \frac{f_{max}}{g}\ \frac{\sigma_v}{\sigma'_v}\ rd$$

where: $\dfrac{\tau_{av}}{\sigma'_v}$ =   induced cyclic stress ratio

f_{max} =   maximum acceleration of ground surface
g =   acceleration of gravity
σ_v =   total vertical stress from the seafloor
σ'_v =   effective vertical stress
rd =   a factor reducing from unity at the surface
to 0.9 at 9 m, 0.75 at 15 m and 0.5 at 30-m
penetration

Shown on Figure 3 are cone point resistance profiles that would indicate possible liquefaction of a fine sand for the given surface ground accelerations. Typical cone point resistance values from three North Sea sites are given on Figure 4. Comparison of Figures 3 and 4 shows that generally the North Sea sands are unlikely to liquefy for ground accelerations upto at least 30 percent of gravity.

Gradation curves of soils susceptible to free field liquefaction (15), (16) are given on Figure 5. Zones where measured cone point resistance suggest potentially liquefiable soil are identified on Figure 4. Particle size distribution of the soils within these zones are shown on Figure 6, and comparison with data presented on Figure 5 indicate that these soils are probably too fine grained to susceptible to free field liquefaction during small seismic activity.

For normally consolidated fine sand relative density can be evaluated from cone point resistance measurements (14) as shown on Figure 7. Similar cone resistance profiles may be developed for overconsolidated sands (14). Comparison of Figures 3 and 7 indicate that dense sands are stable. The procedure used to develop Figure 3 is probably conservative and so medium dense sands are also likely to be stable for the ground acceleration range noted. Loose sands are relatively susceptible to liquefaction, but clean loose sands are not usually encountered in the North Sea.

Slope Stability

Preliminary assessment of infinite slope stability can be evaluated with total stress analysis (2), or in terms of effective stresses (17). More rigorous analyses are available that account for pore pressure changes and are used for detailed site response analyses during design. From total stress analysis the slope should be stable if:

$$s_u/p' > \frac{\text{Sin } 2\beta}{2} + \frac{f}{g} \frac{\gamma}{\gamma'} \text{ Cos}^2 \beta$$

where: s_u = undrained shear strength
        p' = vertical effective pressure
        f = horizontal ground aceleration
        g = gravity
        γ = bulk density
        γ' = submerged unit weight
        β = slope inclination

For clays where the increase in undrained shear strength with penetration is expressed in terms of the ratio S_u/p', the influence of horizontal ground acceleration on stable slope angle is as shown on Figure 8. Typically, regional slopes on the plateau of the northern North Sea are less than 0.5 degrees and may be less than 0.1 degrees (10). In the Central North Sea Basin normally consolidated clays occur at the seafloor (18), (19). In this area clays are generally of low to medium plasticity, Plasticity Index, PI, about 20 to 40, and the ratio S_u/p' about 0.2 to 0.25. For this ratio, horizontal ground acceleration of about 10 percent g may cause slope instability on a 0.5 degree slope. Since there appears to be no evidence of slope failures on the North Sea shelf it may be inferred that seismic activity has not developed ground accelerations in excess of about 10 percent g.

On the continental slopes where inclination is in the order of 2 degrees regionally, lesser ground accelerations are required to induce slope failure. There may be evidence of such failures (9), (10). As indicated on Figure 8, normally consolidated clays are less susceptible to slope failures as plasticity increases. For several sites investigated in the Norwegian Trench moderate to highly plastic clays have been recovered near seafloor with $S_u/p' = 0.4$ to 0.6. Such clays should be stable on shallow slopes for ground accelerations of about 20 percent g.

CONCLUDING COMMENTS

Seafloor soils on the North Sea continental shelf are predominantly overconsolidated clays or sands densified by wave action and ice loading, and appear inherently stable with respect to seismic activity. There is evidence to suggest that slope failures have occurred in the Norwegian Trench and the Norwegian Sea on slope angles of 1 to 2°. These areas should be investigated in detail and mechanisms of failure evaluated.

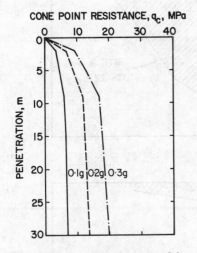

Fig. 3 Cone resistance profiles
for liquefiable sand

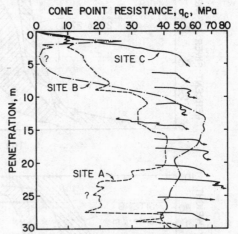

Fig. 4 Typical North Sea cone
resistance profiles

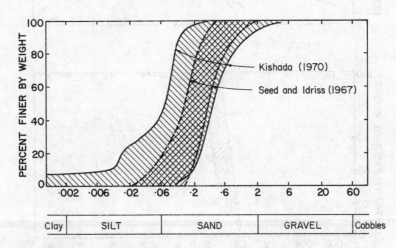

Fig. 5 Gradation of soil susceptible to liquefaction

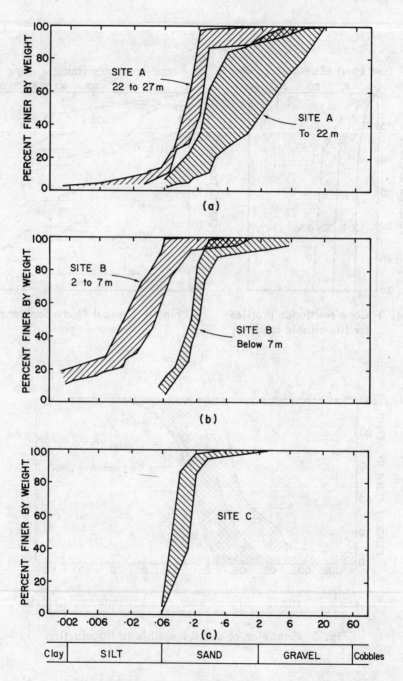

Fig. 6 Typical North Sea grain size curves

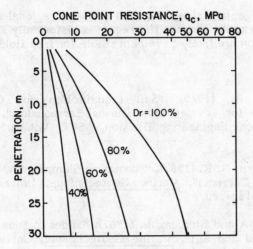

Fig. 7 Interpretation of relative density for normally consolidated sand. (After Schmertmann, 1978)

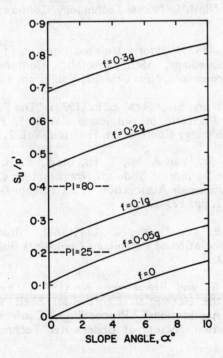

Fig. 8 Ground acceleration to cause infinite slope failure

Geophysics and geology should be used to assess regional stability in areas where seismic activity occurs. If a region is historically stable, then sites within that region are likely to remain stable for free field conditions.

REFERENCES

(1)   Seed, H. B., (1979), "Soil Liquefaction and Cyclic Mobility Evaluation for Level Ground During Earthquakes", Journal of the Geotechnical Engineering Division, ASCE, Vol. 105, No. GT2, pp. 201-255.

(2)   Morgenstern, N. R. (1967), "Submarine Slumping and the Initiation of Turbidity Currents", Marine Geotechnique, University of Illinois Press, pp. 189-220.

(3)   Andersen, A and Bjerrum, L. (1967), "Slides in Subaqueous Slopes in Loose Sand and Silt", Marine Geotechnique, University of Illinois Press, pp. 221-239.

(4)   Demars, K. R., Nacci, V. A. and Wang, W. D. (1977), "Pipeline Failure: A Need for Improved Analyses and Site Surveys", Proceedings, Ninth Offshore Technology Conference, Houston, Vol. 4, pp. 63-70.

(5)   Reifel, M. D. (1979), "Storm Related Damage to Pipelines, Gulf of Mexico", Proceedings, ASCE Speciality Conference, Pipelines in Adverse Environments, New Orleans, Vol. 1, pp. 169-196.

(6)   Sterling, G. H. and Strohbeck, E. E. (1973), "The Failure of the South Pass 70 "B" Platform in Hurricane Camille", Proceedings, Fifth Offshore Technology Conference, Houston, Vol. 2, pp. 719-730.

(7)   Moore, T. C. Jr., Van Andel, T. H., Blow, W. H. and Heath, G. R. (1970), "Large Submarine Slide off Northeastern Continental Margin of Brazil", American Association of Petroleum Geologists Bulletin, Vol. 54, No. 1, pp. 125-128.

(8)   Heezen, B. E. and Drake, C. L. (1964), "Grand Banks Slump", American Association of Petroleum Geologists Bulletin, Vol. 48, No. 2, pp. 221-233.

(9)   Gunleiksrud, T. and Rokoengen, K. (1979), "Regional Geological Mapping of the Norwegian Continental Shelf with Examples of Engineering Applications", Proceedings, Conference on Offshore Site Investigation, Society of Underwater Technology, London, pp. 23-35.

(10) Fannin, N. G. T. (1979), "The Use of Regional Geological Surveys in the North Sea and Adjacent Areas in the Recognition of Offshore Hazards", Proceedings, Conference on Offshore Site Investigation, Society of Underwater Technology, London, pp. 5-22.

(11) Seed, H. B., Martin, P. P. and Lysmer, J. (1976), "Pore Water Pressure Changes During Soil Liquefaction", Journal of the Geotechnical Engineering Division, ASCE, Vol. 102, No. GT4, pp. 323-346.

(12) Idriss, I. M., Dobry, R., Doyle, E. H. and Singh, R. D. (1976), "Behaviour of Soft Clays Under Earthquake Loading Conditions", Proceedings, Eighth Offshore Technology Conference, Houston, Vol. 3, pp. 605-616.

(13) ASTM (1978), "Penetration Test and Split-Barrel Sampling of Soils", Annual Book of ASTM Standards - Part 19, D 1586-67 (1974), American Society of Testing and Materials, pp. 235-240.

(14) Schmertmann, J. H. (1978), "Guidelines for Cone Penetration Test Performance and Design", U.S. Department of Transport Publication No. FHWA-TS-209, July, 143 pp.

(15) Kishada, H. (1970), "Characteristics of Liquefaction of Level Sandy Ground During the Tokachioki Earthquake", Soils and Foundations, Vol. 10, No. 2, pp. 103-111.

(16) Seed, H. B. and Idriss, I. M. (1967), "An Analysis of the Soil Liquefaction in the Niigata Earthquake", Journal of the Soil Mechanics and Foundation Division, ASCE, Vol. 92, No. SM2, pp. 53-78.

(17) Hampton, M. A., Bouma, A. H., Carlson, P. R., Molnia, B. F., Clukey, E. C. and Sangrey, D. A. (1978), "Quantitative Study of Slope Instability in the Gulf of Alaska", Proceedings, Tenth Offshore Technology Conference, Houston, Vol. 4, pp. 2307-2318.

(18) Duvivier, S. and Henstock, P. L. (1979), "Installation of the Piled Foundations and Production Modules on Occidental's Piper A Platform", Proceedings, Institution of Civil Engineers, Part 1, Vol. 66, pp. 407-436.

(19) Sutton, V. J. R., Rigden, W. J., James, E. L., St. John, H. D. and Poskitt, T. J. (1979), "A Full Scale Instrumented Pile Test in the North Sea", Proceedings, Eleventh Offshore Technology Conference, Houston, Vol. 2, pp. 1117-1133.

SEISMOLOGY : A GEOTECHNICAL ENGINEER'S VIEW POINT

Y. ZACZEK
Civil engineer and geotechnical engineer

Leader of the geotechnical group
Electrobel International Engineering

ABSTRACT

This paper is intended to present the interpretations the
geotechnical engineer has to produce from the seismological
data.

Buildings fall into two categories when parasismic engineering is concerned : normal buildings to which parasismic
codes are directly applicable and constructions which
require specific computations because of their complexity
or their high safety level.

The French PS 69 and the American UBC codes are briefly
described.

The risk of liquefaction and the phenomenon are explained.
Some problems of offshore construction are provided.

The final section presents the criteria of nuclear power
plant siting according to US regulations followed in
Belgium, particularly the procedures for the determination
of the Safe Shutdown Earthquake and the Operating Basis
Earthquake.

261

A. R. Ritsema and A. Gürpınar (eds.), Seismicity and Seismic Risk in the Offshore North Sea Area, 261–288.
Copyright © 1983 by D. Reidel Publishing Company.

1. INTRODUCTION

Before the seventies, the engineer in geotechnics in our
countries was only concerned by the influence of the soil
directly surrounding the structures, generally located in
Belgium where the seismicity level is low. The geotechnical
engineer should provide the civil work engineer with the
bearing capacity of the soil layers, the settlement or the
heaving of buildings, the eventual presence of karst or dangerous soil like gypsum, montmorillonite, etc...

Presently the constructions are becoming larger and larger
and, especially when nuclear power plants are concerned,
soil mechanics are becoming far more important. He has
therefore to give also the seismological data needed for
the structural design of buildings. But the problem consists in establishing the link between seismologists and
engineers. The seismologist speaks about magnitude when
the engineer prefers to consider the acceleration of the
ground at the site.

For the particular case of Belgium, the authorities decided
to follow the American regulations as regards the construction of our nuclear power plants. This fact started an
intensive collaboration between ELECTROBEL and Mr VAN GILS,
leader of the section of seismology at the "Observatoire
Royal de Belgique" at Uccle. A new study of the Belgian
earthquakes was carried out and new isoseismic maps were
drawn as a result of this study (1). Nowadays, the seismic
data needed to investigate new sites for nuclear power
plants in our country can be found in the study mentioned
above.

When new projects are to be elaborated, even for sites located overseas having a higher level of seismicity, the
geotechnical engineer must face specific problems concerning the dynamics of the soil. He has to estimate the
risk of liquefaction of the foundation soil when it consists of granular and saturated soil affected by heavy
earthquakes. This phenomenon will be explained hereafter.

Specific considerations to our offshore constructions shall
be presented in a particular chapter. Regarding the new
harbour and liquid gas terminal at Zeebrugge (Belgium),
actually under construction, the offshore and the onshore
seismic effects must be taken into account.

2. TWO GROUPS OF CONSTRUCTION DESIGNS

From the parasismic point of view, there are two different
groups of data asked to the Belgian geotechnical engineer.
On the one hand, we find standard buildings such as hospi-
tals, industries, bridges, etc..., on the other hand we have
the constructions requiring a high safety level, specific
computations of displacements, response spectra at each level, etc... e.g. nuclear power plants, liquid gas storages,
barge elevator (in Ronquières, Belgium), large dams, etc...

2.1. Standard Buildings

In Belgium, we are not building such constructions following
parasismic rules, unless the owner specifically requires
it. In that case, we use codes such as the French Code PS 69
or equivalent ones. When doing so the dynamic forces are
replaced by equivalent static loads.

As we start to work intensively in overseas countries, some
of them having a high degree of seismicity, we must take
into account local parasismic rules and codes or any other
well known codes.

These codes need civil work data such as the type of structure, the description of foundation or type of soil, and
the sismic data. The latter data are the most difficult
to be obtained by the engineer, because they consist for
the greater part in intensities either in the MSK or in the
MM scale as well as in the ground acceleration at the site
to which the buildings have to withstand. This difficulty
also is increased by the fact that our last projects of construction are located in arid or semi-arid regions with very
small populations. This means that traces of felt or observed earthquakes are very scarce and increases the difficulty
in establishing the isoseismal maps or even the maximum observed intensity in the region.

In general for standard buildings, we can obtain the intensity from a local seismological centre or from some correlations between the intensity and the magnitude given by
international centres (CSEM in France, IGS or ISS in UK
...). In some countries a zoning map, drawn to American
UBC code, is available so that we can directly apply this
code and find the correlated dynamic loads.

The use of these codes is detailed hereafter (Section 3).

2.2. Construction with Specific Requirements

Equivalent static loads given by the codes do not adequately simulate the reality during an earthquake. In such cases, we have to make dynamic computations.

These computations are based on the frequency description of the reference earthquake consisting of a given free field response spectrum applicable to the site. This response spectrum is the representation for various damping coefficients, generally on a trilogarithmic paper, scaling the acceleration, the velocity and the displacement versus the frequency of the free field soil at the site during the earthquake to which the construction has to resist. Each earthquake has a different response spectrum and some people have plotted the curve containing several spectra of earthquakes of the region. NEWMARK, BLUME and KANWAR produced the NBK design response spectrum (2) used by the US Atomic Energy Commission for the Regulatory Guide 1.60.

These computations also need geodynamic data. The ground surrounding the foundations has to be described in such a way that the dynamic modulus of elasticity or the shear modulus and the damping coefficient can be determined for each layer of the soil.

Performing seismic refraction tests directly on the proposed site for the new construction, the geotechnical engineer determines the velocities of longitudinal (or compression) and transversal (or shear) waves. The dynamic elasticity modulus (Ed) or shear modulus (Gd) can be obtained with formula 2.1-b, c. In the case of hard soil and rock, we can measure the longitudinal (V_L) and the transversal (V_T) velocities together. The moduli Ed and Gd and the Poisson ratio Ud may be derived from formulas 2.1.-a.

$$U_d = \frac{(1 - 2\,R^2)}{2\,(1 - R^2)} \qquad\qquad 2.1.-a$$

$$E_d = \rho\,V_L^2\,\frac{(1 + U_d)\,(1 - 2\,U_d)}{(1 - U_d)} \qquad\qquad 2.1.-b$$

$$G_d = \rho \times \overline{V}_T^{\,2} \qquad\qquad 2.1.-c$$

if $\quad R = V_T/V_L$

where V_L = velocity of compression waves

V_T = velocity of shear waves

ρ = mass density (specific weight divided by the acceleration of gravity).

In the case of soil, the transversal velocity is very difficult to be measured. The values of the moduli have therefore to be computed taking by into account the measured V_L and a supposed value for the dynamic Poisson ratio, or by considering the measured V_L and a value of V_T provided by the literature.

The shear wave velocities derived from seismic refraction tests provide the elastic behaviour of soil/rock at small shear strains in the order of 10^{-4} per cent. However, as soils and rocks are non-linear materials, the shear modulus decreases with increasing strain amplitude, i.e. from the relatively small strain levels associated with a seismic test to the higher strain levels associated with macro-earthquakes. Similarly, the material damping, which occurs through minute slips of soil particles during straining, increases with shear strain.

The variation of modulus and damping with shear strain can be determined in the laboratory in cases where good samples which are representative of in-situ conditions are available. To obtain the strain dependent soil characteristics of shear modulus and material damping in the laboratory, we use the resonant column test, the shear tests or the torsional shear tests conducted under cyclic loading procedure. Alternatively, the results of research programs on similar materials may be used to establish the relationships on a statistical basis.

Values of the shear modulus and of the damping ratio are given in a study of BOLTON-SEED and IDRISS (3, 4, 5) for sand, satured clays and in a limited way for gravelly soils and peats.

3. THE PARASISMIC CODES

3.1. Utilization

During an earthquake, the soil is strongly shaken, inducing forces applied to the structure in any direction. For common constructions to which a parasismic code can be applied, this dynamic action is represented by equivalent static load in the horizontal and the vertical directions.

Around the world, only about thirty countries have or have
consistently taken into account parasismic rules, despite
of the number of countries with high seismicity. For example, we can mention the French PS 69 (6), the American UBC
(7), the Portuguese (8) or Algerian codes (9), etc...

When we find no accurate local code, we use the PS 69 and
the UBC for overseas countries (Saudi Arabia, United Arab
Emirates, Libya, Iraq, etc...). Short descriptions of PS 69
and UBC codes are given hereafter.

3.2. Short Descriptions

3.2.1. The French Code PS 69. After the presentation
of the aim of this code, the authors point out the strong
importance of the design and the use of adapted materials
resistant to dynamic forces. They advise the engineers
against designing cantilever or asymetric wind bracing, and
provide recommendations for the foundations, the structures
and special rules applicable to the various materials used
and to construction methods. The degree of protection
conferred on a building by applying these rules on the
basis of the given nominal intensity i_N is a function of
the definition and the use of this building (Figure 1).

The equivalent static load is a direct function of
the weight (formula 3.2.1.) by using the multiplication of
four factors :

$$\overline{S} \quad = \quad \alpha \ \beta \ \gamma \ \delta \ \overline{W} \qquad\qquad (3.2.1.)$$

where $\overline{S}$ equivalent static vector

$\overline{W}$ weight vector.

In this formula α, β, γ, δ, are coefficients without dimension, named respectively :

- intensity coefficient
- response coefficient
- distribution coefficient
- foundation coefficient.

The intensity coefficient α makes possible the adjustment of the resistance of a construction to the seismic intensity against which the designer or the collectivity expect to protect it. It depends on the nominal intensity

GROUPS	CONSTRUCTIONS		Nominal intensity i_N		
	DEFINITION	EXAMPLES	Zone I	Zone 2	Zone 3
I	Buildings which present a so-called normal risk for the population	Dwellings, offices, factories, workshops	7 (α = 0,5)	8 (α = 1)	8,6 (α = 1,5)
II	Buildings presenting a special risk owing to the fact that they are frequently visited or because they are of prime importance to the livelihood of a region	Schools, stadiums, entertainment theatres, passenger halls, thermal power plants, etc..	7,6 (α = 0,75)	8,3 (α = 1,2)	8,8 (α = 1,7)
III	Works for which the safety is of prime importance where Civil Defence requirements are concerned.	Hospitals, barracks	8 (α = 1)	8,6 (α = 1,5)	9 (α = 2)
IV	Works for which disorganization would present a particularly severe risk.	Some installations related to the use of atomic energy	to be examined for each case		

Values of the nominal intensity i_N according to the seismicity of zone 1 (low), zone 2 (medium) and zone 3 (strong)

(French PS 69 – 1976)

FIGURE 1

i$_N$ for which the project must be established. The values
of α corresponding to nominal intensities are read on the
functional scale on Figure 2a.

The response coefficient β characterizes the importance of the structural response to a tremor having an equal
intensity to that of the reference intensity. It depends
on :

- the period T of the fundamental vibration mode of the construction in the direction studied

- the degree of damping of the work

- as a secondary consideration, the nature of the foundation
  stratum.

This coefficient is computed with formulae and plotted on
Figure 2b against the period T taking into account the type
of soil and building.

The distribution factor γ depends only on the structure and characterizes the behaviour of each elementary
mass, to which the seismic load is related, in comparison
with the total mass of the building. In the contemporary
type of framed structure, we may consider, unless there is
an anomaly in the distribution of loads, that all the masses
are concentrated at floor levels. In the case of residential
buildings comprising a series of what might be considered
identical storey heights, may be expressed in terms of
the floor rate r as calculated from the base. If n denotes
the number of floors, the factor applicable to the floor
rate r is :

$$\gamma_r = \frac{3r}{2n + 1}$$

See Figure 2c.

The foundation factor δ, which is independent of the
dynamic properties of the building, is a corrective factor
which takes into account the incidence of foundation conditions on the behaviour of the structure (Figure 2d).

3.2.2. The Uniform Building Code. This code, being
only a part of a general code of construction, is not as
detailed as the French one, but both proceed in the same
way. Some abstracts of this code are presented hereafter.

α	0,5	1	2	3	4	5
Intensité nominale	6	7	8	9	10	
Degré macro sismique	VII	VIII	IX	X	XI

FIGURE 2a

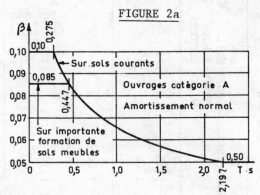

FIGURE 2b

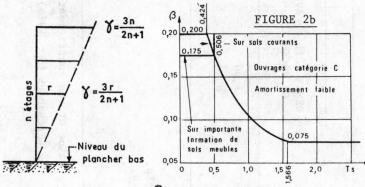

FIGURE 2c

Values of
coefficients

α, β, γ
(French PS 69–1976)

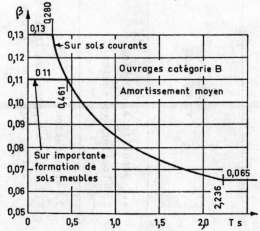

METHOD OF FOUNDATION	S O I L			
	A DENSE SOIL OR ROCK	B AVERAGE SOIL	C SOFT SOIL, RATHER HIGH WATER CONTENT	D WATER-SATURATED MUD AND SILT
1. Superficial footings	1,00	1,15	1,25	
2. Footings on shafts	0,90	1,00	1,15	
3. Foundation raft		1,00	1,10	1,20
4. Piles supported at their tip, driven through soil types B, C or D.		1,10	1,15	1,30
5. Floating piles		1,10	1,30	

Value of the coefficient δ

(French PS 69 – 1976)

FIGURE 2d

Every structure shall be designed and constructed to resist minimum total lateral seismic forces assumed to act nonconcurrently in the direction of each of the main axes of the structure in accordance with the formula 3.2.2.-a

$$V = ZIKCSW \qquad\qquad (3.2.2.-a)$$

where

Z =   Numerical coefficient dependent upon the zone as determined in the zoning maps of the studied country (if existing). For locations in Zone No. 1, $Z = 3/16$. For locations in Zone No. 2, $Z = 3/8$. For locations in Zone No. 3, $Z = 3/4$. For locations in Zone No. 4, $Z = 1$.

I =   Occupancy Importance Factor as set forth in Table No.23 K of the code

K =   Numerical coefficient as set forth in Table No. 23-1 of the code.

C =   Numerical coefficient as specified in Section 2312 (d) of the Code

S =   Numerical coefficient for site-structure resonance

V =   The total lateral force or shear at the base

W =   The total dead load.

The Tables I and K are presented on the Figure 3. The value of C ans S are as indicated hereafter except that the product of CS need not exceed 0.14.

The value of C shall be determined in accordance with the formula 3.2.2.-b. The value of C need not exceed 0.12. The period T shall be established using the structural properties and deformational characteristics of the resisting elements in a properly substantiated analysis. In the absence of a determination as indicated above, the value of T for buildings may be determined by the formula 3.2.2.-c.

Or in buildings in which the lateral force-resisting system consists of ductile moment-resisting space frames capable of resisting 100 per cent of the required lateral forces

TABLE NO. 23-K—VALUES FOR OCCUPANCY IMPORTANCE FACTOR I

TYPE OF OCCUPANCY	I
Essential facilities[1]	1.5
Any building where the primary occupancy is for assembly use for more than 300 persons (in one room)	1.25
All others	1.0

[1]See Section 2312 (k) for definition and additional requirements for essential facilities.

TABLE NO. 23-I—HORIZONTAL FORCE FACTOR K FOR BUILDINGS OR OTHER STRUCTURES[1]

TYPE OR ARRANGEMENT OF RESISTING ELEMENTS	VALUE[2] OF K
1. All building framing systems except as hereinafter classified	1.00
2. Buildings with a box system as specified in Section 2312 (b) **EXCEPTION:** Buildings not more than three stories in height with stud wall framing and using plywood horizontal diaphragms and plywood vertical shear panels for the lateral force system may use $K = 1.0$.	1.33

3. Buildings with a dual bracing system consisting of a ductile moment-resisting space frame and shear walls or braced frames using the following design criteria:
   a. The frames and shear walls or braced frames shall resist the total lateral force in accordance with their relative rigidities considering the interaction of the shear walls and frames
   b. The shear walls or braced frames acting independently of the ductile moment-resisting portions of the space frame shall resist the total required lateral forces
   c. The ductile moment-resisting space frame shall have the capacity to resist not less than 25 percent of the required lateral force
	0.80

4. Buildings with a ductile moment-resisting space frame designed in accordance with the following criteria: The ductile moment-resisting space frame shall have the capacity to resist the total required lateral force	0.67
5. Elevated tanks plus full contents, on four or more cross-braced legs and not supported by a building	2.5[3]
6. Structures other than buildings and other than those set forth in Table No. 23-J	2.00

[1]Where wind load as specified in Section 2311 would produce higher stresses, this load shall be used in lieu of the loads resulting from earthquake forces.

[2]See Figures Nos. 1, 2 and 3 in this chapter and definition of Z as specified in Section 2312 (c).

[3]The minimum value of KC shall be 0.12 and the maximum value of KC need not exceed 0.25.

   The tower shall be designed for an accidental torsion of 5 percent as specified in Section 2312 (e) 4. Elevated tanks which are supported by buildings or do not conform to type or arrangement of supporting elements as described above shall be designed in accordance with Section 2312 (g) using $C_p = .3$.

FIGURE 3   UBC code (1982)

and such system is not enclosed by or adjoined by more
rigid elements tending to prevent the frame from resisting
lateral forces by the formula 3.2.2.-d

$$T = \frac{0.05h_n}{\sqrt{D}} \qquad (3.2.2.-c)$$

$$C = \frac{1}{15\sqrt{T}} \qquad (3.2.2.-b)$$

$$T = 0.10 \, N \qquad (3.2.2.-d)$$

where

D = The dimension of the structure, in feet, in a direction parallel to the applied forces

N = The total number of stories above the base to level
n.

The value of S shall be determined by the formulas
3.2.2.-a, f, but shall be not less than 1.0 or equal to 1.5
if T_s is not properly established

- for T/T_s = 1.0 or less, $S = 1.0 + \frac{T}{T_s} - 0.5 \, [\frac{T}{T_s}]^2$ (3.2.2.-c)

- for T/T_s, greater than 1.0 or less,

$$S = 1.2 + 0.6 \frac{T}{T_s} - 0.3 \, [\frac{T}{T_s}]^2 \qquad (3.2.2.-b)$$

where

T = Fundamental elastic period of vibration of the building or structure in seconds in the direction under
consideration

T_s = Characteristic site period.

After the presentation of this general procedure, the
UBC code gives particular instruction about the lateral forces
on parts or portions of structures, the drift and building
separations, specific structural systems and essential facilities.

3.3. New Developments. During the recent earthquakes
(Friuli 1976, El Ahsnam 1980, etc...) some contradictions
against well established assumptions are appearing, because
the study of destructions becomes more and more accurate
and complete and also because instrumentation is now better
distributed around the world. Among this new information,
we have to point out two major facts.

The first concerns the vertical acceleration that was
taken as one third of the horizontal acceleration during
an earthquake or even ignored. The value of this horizontal
acceleration measured in some buildings is very high, up
to 0.6 g in some places. The study of some damaged buildings shows that, without any horizontal motion, they
collapsed by rupture of overstressed concrete columns.
This last example was presented by Prof. AMBRASEYS for the
earthquake of El Ahsnam.

The second puts the geotechnical engineer very per-
plexed and embarrassed; it consists in the relation between
the acceleration and the intensity of an earthquake. One
of the major input of geodynamics for the parasismic computations is the maximum acceleration at the site. With the
coming of nuclear safety regulations, more and more authors
are publishing different correlations of peak acceleration
with seismic intensity scales. We may say that each author
can find a different correlation based on well improved recordings of earthquakes. We may mention as examples :

    Ishimoto 1932
    Gutenberg, Richter 1942 (10)
    Kawasumi 1951
    Neumann 1954
    Savarensky, Kirnos 1955
    Richter 1958 (11)
    Medvedev, Sponheuer, Karnik 1964 (12)
    Trifumac, Brady 1975 (13)
    etc...

Furthermore the scatter of the correlation acceleration-
intensity becomes wider and wider without taking into consideration the "facility" for the engineer to "detect" the
acceleration from a given magnitude.

The conclusions of these new developments are of two
aspects. For the immediate one, some codes are revalued
(the Algerian code) or are under revaluation (the French
PS 69 code). Philosophical conclusion consists in looking

for new input data more accurate than the acceleration and
maybe in considering these correlations to be only applicable
to the region where the earthquakes, taken to establish the
correlation, occurred. When doing so, we could find a correlation of intensity with acceleration, velocity and displacement for each region of the world.

4. THE LIQUEFACTION OF SOIL DURING AN EARTHQUAKE

4.1. The phenomenon

All the water saturated soils could theoretically be liquefied
when shaken by a very strong vibration. The seismic level
of recorded earthquakes in Belgium is so low that liquefaction could affect only the loose saturated sand.

Its visible consequence at the ground level consists in col-
lapsing soil in crater with mud spouts, in seeping of water,
and in creating areas of quicksand. When a foundation lays
on a soil layer that liquefies, its bearing capacity is sup-
pressed, suddenly but not everywhere and not at the same
time under all the foundations, so the building may sink
generally with tilting. A lot of examples can be found in
Japan.

This phenomenon, very simple to understand, occurs only in
granular soil such as sands and not in clays, because their
permeability is too low to permit the development of extra
pore pressure with the same velocity as a seismic wave. The
granular soil consists of grains, water and air. The action
of this air is neglected hereafter because of its compressi-
bility and its low concentration in a saturated soil. The
stability of a soil, expressed in terms of bearing capacity
and settlement under a load, is based on the friction
between the grains under the load of the overlaying soil
and a given water pore pressure induced mainly by the level
of the water table. This equilibrium of the soil is
expressed in terms of overburden pressure at any depth.
In loose sand during an earthquake, the vibrations compacting the sand induce a sudden extra pore pressure. The
high permeability of this type of material may transmit
this pressure very quickly to the uncompressible water.
The sand is loose and the friction between grains is not
very strong; when the extra pore pressure reaches the
overburden pressure, the friction between grains is
reduced to zero and the grains are floating. The soil has
now the same properties as a liquid having a higher density.

The liquefaction may occur at any depth and may propagate
in any direction where the conditions of density and
seismic energy are suitable. After starting at one point
in the soil, the transmission of the extra pore pressure
extends the liquefied area, until it becomes insufficient
compared to the overburden pressure.

According to BOLTON-SEED and IDRISS (15), the possibility
for a soil to be liquefied depends primarily on the soil
nature, its density, its void ratio, its overburden pressure,
the duration and the intensity of the earthquake.

4.2. The Quantification of the Risk

Mostly in the United States of America and in Japan, different kinds of tests and several methods of computation
try to provide assessments on the liquefaction risk for a
given soil subjected to a known earthquake.

Using programs like SHAKE, we compute the pressure induced
by a vibration at any level of a soil column, and, comparing
this with the overburden pressure, we can estimate the risk.

Performing laboratory tests like the shear dynamic test,
the shaking table etc..., we measure the ability of a soil
to be densified when submitted to a vibrating energy with
the same amplitude as an earthquake. The critical void
ratio and the critical density characterize the state of a
given soil, the volume of which decreases with the vibrations
inducing then extra pore pressure. Correlations between the
grain size distribution of a soil and the liquefaction possibility were published and are based on historical liquefied
areas and on results of laboratory tests. The Figure 4 presents the results of studies by SHANNON and WILSON (14).

The above criteria need density measurements performed directly on undisturbed samples, taken in the studied layers
or indirectly by the use of in situ tests like the Standard
Penetration Test. To obtain undisturbed samples from loose
saturated sand layers, the most liquefiable soils, is a
nearly impossible performance, whereas in situ tests provide
more accurate results in such soils. Taking into account
the results of SPT tests hereabove mentioned, BOLTON-SEED
and IDRISS (4, 5) provide us with minimum quantities of blows
per foot above which no liquefaction could occur during
several earthquake accelerations (Figure 5).

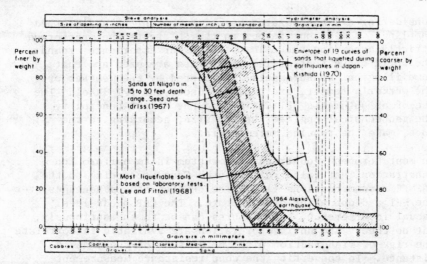

FIGURE 4   The influence of the grainsize distribution on the liquefaction risk (14)

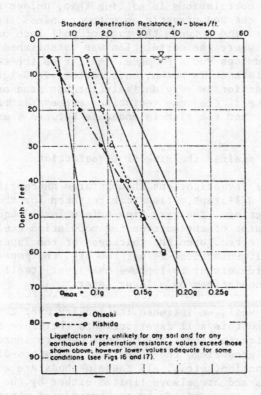

FIGURE 5   For given accelerators, values of the number of blows over which liquefaction is very unlikely to occur.

Considering all the hazards associated with the undisturbed
sampling, we give our favour to the SPT tests when performed
strictly according the ASTM-D 1586. Our experience around
the world shows that different devices are used and that
some illogical measurements, influenced by the weather or
the worker's temper, are provided under the SPT label. The
use of an automatic SPT device and the direct control by
the geotechnical engineer of its size, according to the ASTM
code, make the results more accurate.

In most European countries, the custom is to perform Cone
Penetration Test (also called Dutch Cone) according to the
ISSMFE recommendations (15). The CPT has many advantages over
the SPT and one major disadvantage. The CPT needs less
manual interference, the measures can be taken continously,
but no theory and no interpretation are available to correlate
the risk of liquefaction with the measure of the cone resistance. In converting the cone resistance measurement
into equivalent blows numbers per foot with various correlations, some engineers quote the risk of liquefaction. The
scatter of these correlations is so wide that, unless these
correlations are checked directly on site or unless the great
similarity is verified between the type of soil both on site
and at the place where the correlation was established,
inaccurate conclusions could be made. As far as the relation
density-cone resistance is known, some authors in Belgium
feel that liquefaction is very unlikely for any sand and
for any earthquake if the cone resistance exceeds 8 MPa
(80 kg/cm2), and that the risk is moderate between 4 and
8 MPa.

4.3. The action against the Risk of Liquefaction

When geotechnical investigations show an area susceptible
to liquefaction, different options can be taken for the
proposed construction. The very high safety level required
for the safe-keeping of the surrounding population could
make the site rejected. When the thickness of the layer,
susceptible for liquefaction, is not too big, the geotechnical
engineer can decide either to improve the layer itself or
to support the foundations on a lower layer.

In improving the soil, we increase its density over the critical density, and this soil layer presents therefore no
longer risk of liquefaction. The methods used to improve
the soil are the grounting, the vibration, the vibro-floatation,
the dynamic compaction, etc... All these methods are adapted
to specific cases and are always limited either by the depth
of the layer to be improved, or by the grain size distribution,
or by the chemical content of the soil, etc...

The behaviour of a building during an earthquake, when
foundations are supported on a lower layer covered by a
liquefied layer, is fundamentaly different. During the
earthquake the liquefied soil surrounding the foundations
(e.g. piles) not only removes the soil reaction but also
induces strong horizontal forces on the foundations. This
type of solution therefore requires detailed computations
as indicated in section 2.2.

5. OFFSHORE CONSTRUCTIONS

The Belgian coast is only about 70 kilometres long, and the
territorial waters are not very extensive. Our offshore
problems are therefore limited, and consist mostly in
coastal engineering.

The offshore problems, that our company is currently tackling,
are the construction of a complex of embankments and a
platform erected in the North-sea to realize a new harbour
at Zeebrugge in Belgium, and the realization of water intakes
and outfalls in Libya, Abu Dhabi, etc...

In Zeebrugge, we are building a new harbour and a liquid
gas terminal on a platform advancing 1.5 kilometre in the
sea. This platform (400 m wide) consists of hydraulic
sands backfilled between rock embankments. The thickness
of this backfill is about 16 metres. Zeebrugge is located
in a low seismicity area with an intensity between V and
VI on the MSK scale, and the sand of the backfill is very
compact. The parasismic problems are therefore very small.
The high density of the sand avoids the liquefaction phenomenon and only the stability on the embankments during an
earthquake could be of concern. In fact the extra pore pressure produced by an earthquake could be applied on a potential
sliding curve drawn on cross sections of the embankments
during the study of the slope stability (Figure 6).

An other effect we must take into account for our overseas
constructions is the Tsunami phenomenon. For the moment we
are not yet concerned by constructions in the sea, but many
of our industrial complexes in foreign countries are built
on coast-lines. By designing these buildings at a higher
level than that attained by seismic sea-waves and locating
them, when possible, not directly in front of the sea, we may
protect them against the tsunami phenomenon.

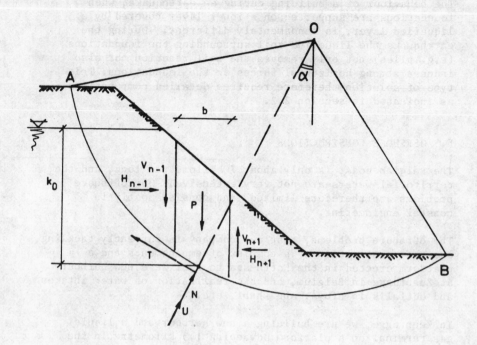

Forces applied to a slice

- H and V : internal forces
- P : weight of the slice
- T : cohesion developed along the sliding line
- N : granular reaction force
- U : hydraulic force

Effect of an earthquake

- Before earthquake U = Uo : pore pressure due to the
$$\text{water tabel } (Uo = ko \times 1T/m3)$$

- During earthquake U = Uo + $\Delta U(t)$ where $\Delta U(t)$ is the extra pore pressure function of the time

FIGURE 6 Stability of slopes : the influence of an earthquake.

6. THE NUCLEAR POWER PLANTS (NPP)

6.1. The US Regulations

Nuclear power plants in Belgium are built according to the US Nuclear Regulations. For the geotechnical aspect, we follow the 10 CFR (Code of Federal Regulations) part 100 entitled "Reactor site criteria", concerning the geology, the seismology and the geotechnics. The 10 CFR 100 is treated as a law and is explained in the Regulatory Guide 1.70. Each nuclear power plant has to present the Preliminary and Final Safety Analysis Report (PSAR and FSAR), that must be approved by experts before the NPP becomes operational. To help the engineer in writing his safety analysis report, standard reviews detail the regulatory guides.

Appendix A of the 10 CFR 100 provides seismic and geologic criteria setting forth the main geological and seismological features to evaluate the suitability of the NPP site. These criteria describe the procedures for determining the vibratory ground motion basis at the site due to earthquakes and for studying the surface faulting to detect the eventual presence of active faults near or under the constructions.

Regulatory guides 1.132 and 1.138 provide us with recommended in situ and laboratory tests and procedures.

6.2. The Main Basis Definitions (abstract from 10 CFR 100-A)

This regulation gives well known definitions, such as magnitude, intensity, fault, etc... Other new definitions were established in 1975, governing all the safety reports :

a. The "Safe Shutdown Earthquake" is that earthquake which is based upon an evaluation of the maximum earthquake potential considering the regional and local geology and seismology and specific characteristics of local subsurface material. It is that earthquake which produces the maximum vibratory ground motion for which certain structures, systems, and components are designed to remain functional.

b. The "Operating Basis Earthquake" is that earthquake which, considering the regional and local geology and seismology and specific characteristics of local subsurface material, could reasonably be expected to affect the plant site during the operating life of the plant; it is that earthquake which produces the vibratory growth motion for which those features of the nuclear power plant necessary for continued operation without undue risk to the health and safety of the public are designed to remain functional.

c. A "capable fault" is a fault which has exhibited one or more of the following characteristics :

(1) Movement at or near the ground surface at least once within the past 35,000 years or movement of a recurring nature within the past 500,000 years.

(2) Macro-seismicity instrumentally determined with records of sufficient precision to demonstrate a direct relationship with the fault.

(3) A structural relationship to a capable fault according to characteristics (1) or (2) of this paragraph such that movement on one could be reasonably expected to be accompanied by movement on the other.

d. A "tectonic province" is a region of the North American continent characterized by a relative consistency of the geologic structural features contained therein.

e. The "control width" of a fault is the maximum width of the zone containing mapped fault traces, including all faults which can be reasonably inferred to have experienced differential movement during Quaternary times and which join or can reasonably be inferred to join the main fault trace, measured within 10 miles along the fault's trend in both directions from the point of nearest approach to the site. (See Figure 7).

All these definitions are used for our NPP, except the notion of tectonic province. This notion is not adaptable to Belgium because of the wide geological and tectonic variations within 320 km of the NPP site, i.e. within the radius of investigation demanded by the PSAR.

6.3. The determination of SSE and OBE

For the NPP of TIHANGE units 2 and 3, we performed, with VAN GILS, a complete reevaluation of the seismicity of Belgium (1), because the regulations require us to investigate, within 320 km of the NPP, the geology, the tectonics, the seismology, etc...

The isoseismal map of Belgium (Figure 8) was obtained during this study, with the location of all the epicentres more than degree III on the MSK scale.

Any part of all the faults within 320 km of the site which may be of significance in determining the SSE, or which may be capable must be investigated.

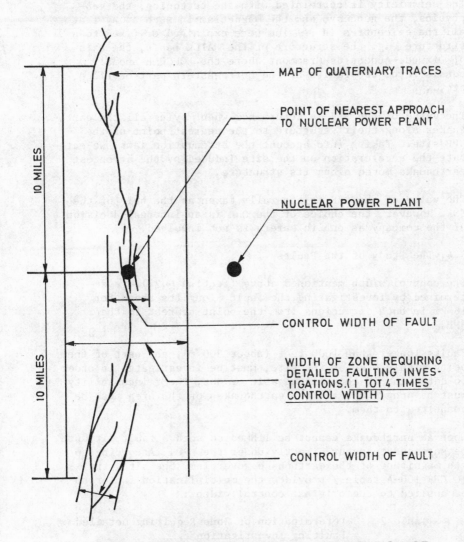

DIAGRAMMATIC ILLUSTRATION OF DELINEATION OF WIDTH OF ZONE REQUIRING DETAILED FAULTING INVESTIGATIONS FOR SPECIFIC NUCLEAR POWER PLANT LOCATION.

FIGURE 7

The seismicity is correlated with the tectonics, the geophysics, the geodesy, and the magnetism in such a way that
all the epicentres in Belgium were explained or fixed to a
structure e.g. the structure of the HAINE basin, the axis
of Dixmude-Audenarde-Grammont where the BOUGUER anomaly
equals zero, the faults at the Northeastern part of Liège,
etc...

The value of the SSE is determined when moving all the earthquakes along their structure to the nearest point of the
NPP site. Taking into account the attenuation laws, we estimate the acceleration on the site induced by the strongest
earthquake moved along its structure.

The value of the OBE is generally taken as the half of the
SSE. However, the choice of the OBE is an internal decision
of the company as public safety is not involved.

6.4. The Study of the Faults

The control width mentioned above (section 6.2.), is determined by investigating the fault along its trend for
16 km in both directions from the point nearest to the
NPP.

Faults longer than 1000 feet (about 300 m), any part of them
being within 8 km of the site, must be investigated in order
to determine if they are capable or not. Their incapability
must be proved and all the earthquakes of the area must be
connected to them.

When an earthquake cannot be linked to such a fault, it must
be supposed to occur directly under the NPP. According to
the magnitude of the earthquake governing the site, the
10 CFR 100-A table 2 provides the multiplication factor to
be applied to the original control width :

Table 2   Determination of Zone Requiring Detailed
               Faulting Investigation

Magnitude of earthquake	Width of zone requiring detailed Faulting investigation (see Figure 1)
Less than 5.5	1 x control width
5.5-6.4	2 x control width
6.5-7.5	3 x control width
Greater than 7.5	4 x control width

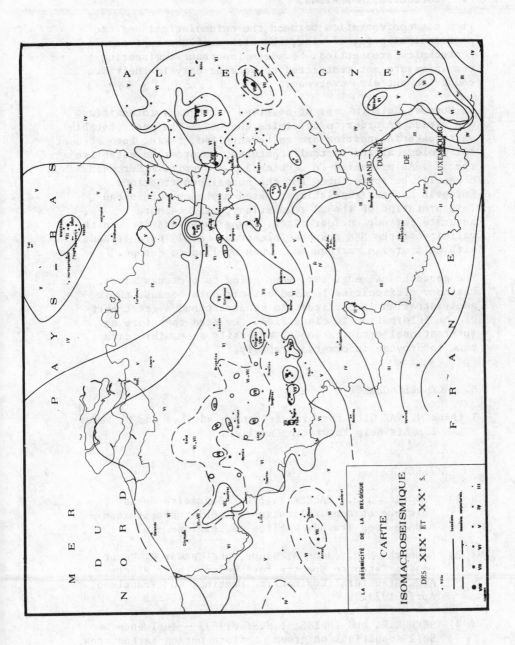

FIGURE 8
Isomacroseismic map of BELGIUM IX and XX centuries.

7. CONCLUSIONS AND WISHES

From the confrontation between the seismological and the
geotechnical point of view, the wishes of the engineer in
geotechnics are settled. Nowadays important engineering
companies are not restricted to working only in their own
countries but also overseas.

In Europe, uniform maps of observed earthquake intensities
are needed in order to determine the OBE. The most probable
earthquake intensities are to be supplied, taking into account
all geological and tectonic features, in order to establish
the SSE. As presented in section 3.3., a uniform European
code will be appreciated for the normal constructions.
European nuclear regulations, based maybe on a European
interpretation of the US regulations, will be more
accurate for our nuclear sites. Many engineers dispute the
validity of the NBK response spectrum, mainly established
with Californian earthquakes, when applied to Europe.

The geotechnical engineer, when going to overseas sites,
has many difficulties to find seismic data because their
administration is difficult to penetrate and a great part
of the information is classified. We need therefore an
international organization or an easily accessible data
bank to provide information rapidly.

8. ACKNOWLEDGEMENT

I thank N. VAN GILS for his discerning advices and M. Peter
NOLAN for his help in writing English.

9. REFERENCES

(1)   VAN GILS J.M., ZACZEK Y. : La séismicité de la
       Belgique et son application en génie parasismique -
       Annales des Travaux Publics de Belgique - n° 6-1978.

(2)   Newmark, N.M., John A. Blume, and Kanwar K. Kapur
       Design Response Spectra for Nuclear Power Plants,
       ASCE Structural Engineering Meeting, San Francisco,
       April 1973.

(3)   SEED H.B. and IDRISS, I.M. : (1969) - Influence of
       Soil Conditions on ground Motions During Earthquakes,
       - Journal of the Soil Mechanics and Foundations Division,
       ASCE, Vol. 95, No SMI, Proc. Paper 6347, January,
       pp. 99-137

(4)   SEED, H.B. and IDRISS, I.M. : (1971) - A simplified
       Procedure for Evaluating Soil Liquefaction Potential
       - Journal of the Soil Mechanics and Foundations Division,
       ASCE, Vol 97, No. SM9, September, pp. 249-274.

(5)   SEED, H.B.and IDRISS, I.M. : (1967) Analysis of Soil
       Liquefaction Niigata Earthquake - Journal of the Soil
       Mechanics and Foundations Division. ASCE, Vol. 93,
       SM3, Proc. Paper 5233, May, pp. 83-108

(6)   France - Règles parasismiques françaises 1969 et
       annexes - Editions Eyrolles 1976.

(7)   US Uniform Building Code - International Conference
       of Building Officials - Whittier, California 90601-1982.

(8)   Portugal - Règlement des sollicitations dans les
       édifices et les ponts - chapitre IX - Diario do Governo
       n° 268 - Journal officiel portugais 18/11/1961.

(9)   Algérie - Recommandations provisoires aux bâtiments
       à édifier dans les régions sujettes aux séismes -
       Ministère des Travaux Publics et de la Construction
       - Direction de la Construction et de l'Habitat.

(10)   GUTENBERG B., RICHTER C.F. : Earthquake Magnitude,
       Intensity, Energy and Acceleration - Bull. Seism.
       Soc. Am. - Vol. 32,3-1942

(11)   RICHTER C.F. : Elementary Seismology - Freeman and
       C° - 1958

(12)   MEDVEDEV S., SPONHEUER W., KARNIK V. : Neue Seismische
       Skala. Veröff. Inst. Bod. Jena - Heft 77 - Akademie
       Verlag - Berlin-1964.

(13)   M.D. TRIFUNAC and A.G. BRADY : "On the correlation
       of seismic intensity scales with the peak of recorded
       strong ground motion" - Bulletin of the Seismological
       Society of America, Vol. 65 No 1, pp.139-162-Feb.1975.

(14)   SHANNON et WILSON : Soil Behavior under Earthquake
       Loading Conditions - Report prepared for U.S.A.E.C.-1971.

(15)   IDRISS I.M. and SEED H.B. : "Effects of local geologic
       and soil conditions on damage potential during earth-
       quakes" - 2nd International Congress of the International
       Association of Engineering Geology - Sao Paulo-1974.

(16) INTERNATIONAL SOCIETY FOR SOIL MECHANICS AND FOUNDATION
      ENGINEERING : "Report of the sub-committee on the pene-
      tration test for use in Europe" - The Secretary General
      ISSMFE, King's College London Strand, WC2R 2LS,
      U.K.-1978.

TESTING FOR LIQUEFACTION POTENTIAL

D.J. Mallard

Generation, Development and Construction Division,
C.E.G.B., Barnwood, Gloucester, U.K.

Field and laboratory tests have been carried out in order to investigate the liquefaction potential of an overconsolidated Plio-pleistocene sand deposit on the coast of the southern North Sea, where due to the fabric of the sand both types of testing were kept as simple as possible. This contribution describes practical lessons which have been learnt with respect to the appropriate techniques for this specific topic of investigation as compared with conventional British geotechnical investigation practice.

INTRODUCTION.

In 1980 the C.E.G.B. conducted its first formal assessment of the liquefaction potential of a site. In doing this work, certain observations were made which may be of interest to other engineers concerned with this topic. The work is relevant to this workshop in that the material tested is a Plio-pleistocene sand at a site on the Suffolk coast in the extreme south of the North Sea.

In spite of the relatively infrequent occurrence of damage caused by liquefaction, it is a subject of major concern and there is now a large body of literature on the subject so that an engineer is faced with a considerable list of apparently significant parameters. Because of the nature of the soil considered in this investigation (see below) and mindful of Professor Peck's (1) timely warning against reliance on over-sophisticated quasi-science, it was decided to keep the approach as simple as possible. Concentration was directed towards straightforward in-situ

A. R. Ritsema and A. Gürpınar (eds.), Seismicity and Seismic Risk in the Offshore North Sea Area, 289–302.
Copyright © 1983 by D. Reidel Publishing Company.

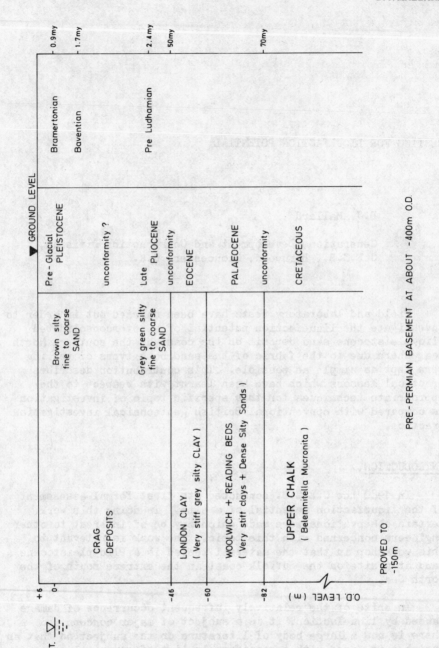

FIG. 1   TYPICAL GEOLOGICAL SECTION

penetrometer testing coupled with basic classification-type laboratory work. By this means, existing published information on liquefaction, both from the small amount of first hand field data and from the large amount of parametric laboratory research, could be utilised to a rational and appropriate extent.

Previous geotechnical investigations of the site provided a large amount of general data which reduced the need for further laboratory testing to a minimum consistent with the resolving of any anomalies.

SITE GEOLOGY.

A typical geological section is shown in fig. 1. This discussion is concerned only with the Crag Deposits and within these strata identification of the various East Anglian stratigraphical stages has been based on pollen and mollusc spectra (2). It is thought from more recent foraminiferal analysis that at least one of the stages apparently missing between the Pre-Ludhamian and the Baventian has been identified (3).

During the penultimate (Wolstonian) glaciation, the Crag would have been covered by a chalky boulder clay which was subsequently removed by post-glacial stream erosion. It therefore should be regarded as an over-consolidated deposit.

The ground water level shown on fig. 1. is typical, but is subject to both tidal and seasonal fluctuations.

PRE-1980 DATA ON CRAG DEPOSITS.

From earlier work, it was apparent that the sands have a marked fabric which in places produces a clearly identifiable structure. These complications arise from their deposition in a shallow-water marine environment being manifested as impersistent horizontal clay layers generally less than 25mm thick and as thin bands of extremely shelley material. Also, generally at depth in the Pre-Ludhamian materials, are found thin cemented bands normally less than 100mm thick.

The material is therefore not ideal for laboratory testing and indeed has virtually defied genuinely undisturbed sampling.

The grading of the Crag is demonstrated in fig. 2. where Effective Grain Size (D_{50}) and Uniformity Coefficient (D_{60}/D_{10}) are plotted against level. Some measure of the incidence of clay lenses and shells is provided by fig. 3.

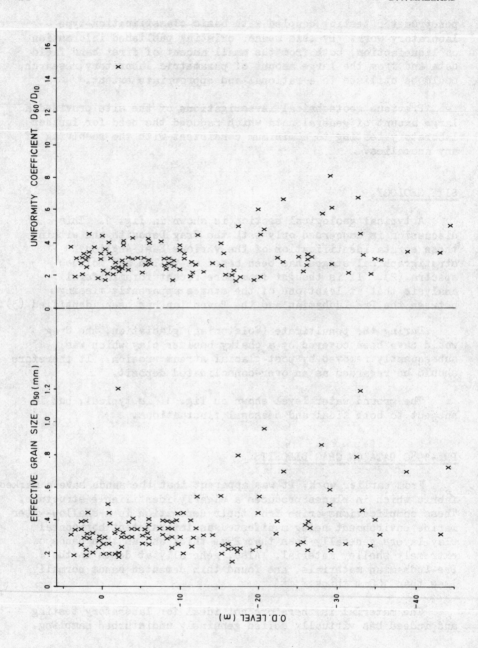

FIG.2   CRAG DEPOSITS GRADING RESULTS

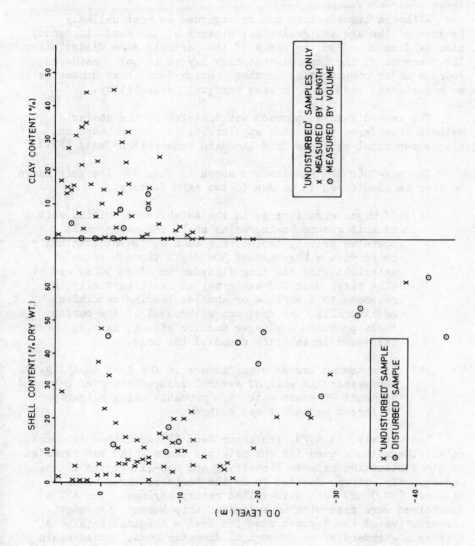

FIG.3 SHELL AND CLAY CONTENT OF CRAG DEPOSITS

FIELD TESTING.

Although liquefaction can be regarded as most unlikely
because of the age and geological history of the sand, it cannot
also be discounted on the basis of the particle size distribution.
The presence of the impersistent clay layers is not considered
to be a major complication in that pumping tests have demonstrated
no significant reduction in mass vertical permeability.

The investigative approach was dictated by the analysis
methods considered to be most applicable, ie, those based on
simple empirical evidence from Standard Penetration Tests (SPT).

The pre-1980 SPT results are shown in fig. 4. The very large
scatter is considered to be due to two main factors:

(a) Of major significance is the established British method
     of soft ground boring using shell and auger cable
     operated gravity tools. For such equipment eventually
     to produce a borehole of 50m depth through granular
     materials, the starting diameter is often 300mm and at
     this size, each SPT measures, at least partially, a
     response to a surface or shallow loading condition.
     Additionally, the frequent withdrawal of the boring
     tools produces a plunger suction effect, causing
     increased instability ahead of the hole.

(b) Of somewhat lesser significance is the fact that fig. 4.
     represents the work of several contractors over a period
     of about 20 years with each probably using slightly
     different equipment and methods.

Fortunately, in 1979, Professor Seed (4) described in some
detail the methods used for the original tests which had resulted
in his distinction between liquefying and non-liquefying sites.
Consequently, it was decided to do the tests in three small
diameter (< 98mm) bentonite-filled rotary borings. The SPT's
themselves were done with an automatic trip hammer (the most
conservative of the devices used for Seed's original data set)
driving a standard spoon through NX diameter rods, chosen again
to introduce an element of conservatism by virtue of their
rigidity. The blow count was taken always for a penetration from
150 to 450mm below the base of the hole.

The results from these tests are shown in fig. 5. and
comparison with fig. 4. shows them not only to demonstrate higher
mean resistance but also much less scatter particularly in the
depth range of interest (down to about -20 m.O.D.). Probably
the increase in scatter at lower levels is due both to the localised
cemented layers and to the increasingly variable loss of energy

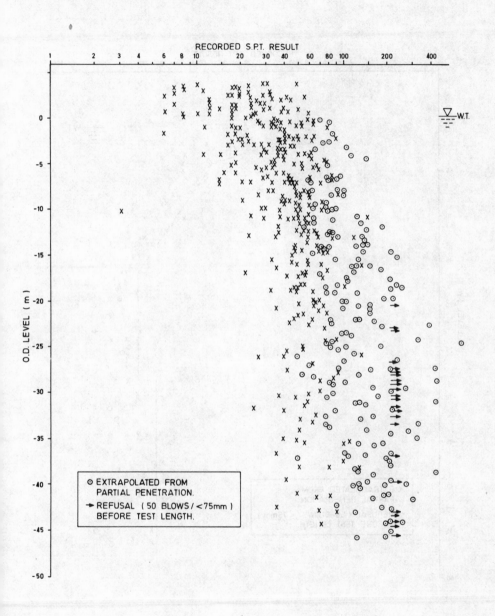

FIG. 4   PRE-1980 S.P.T. RESULTS

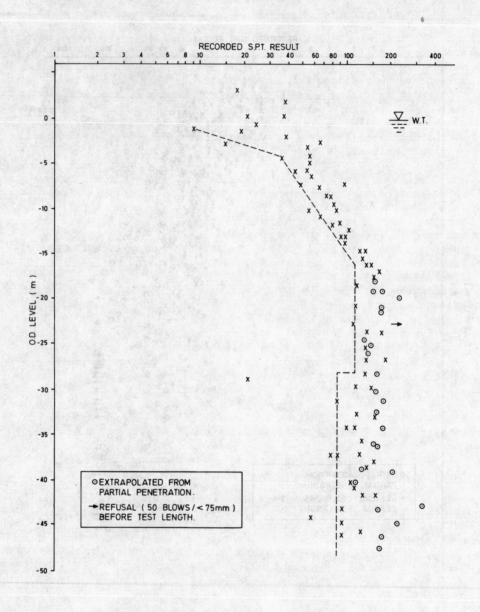

FIG. 5   1980 S.P.T. RESULTS FROM SMALL DIAMETER ROTARY BOREHOLES

in the longer rods.

The lower bound line marked on fig. 4. excludes two obviously erroneous results at depth but includes two probably doubtful values of 9 and 15 just below the water table. These two latter values are judged doubtful on the basis of tests results from 1.5m above and below each of them. They appear to demonstrate that, even with this method of boring and testing, some disturbance close to the water table in unavoidable. However, they are not excluded because they may be representative of conditions in the original empirical liquefaction data.

One major distinction between figs. 4. and 5. is the number of boreholes they represent and to address this possible bias from a small number of holes, two new shell and auger boreholes were put down using the same testing technique and equipment as was used for the fig. 5. results, and the best available boring methods. They started at 250mm diameter, were bored continuously (with no overnight breaks) under a full reverse head of bentonite from a level about 6m above the water table, and the scatter of results covered almost the full range of fig. 4. being particularly pronounced in the depth range of interest.

The three small diameter rotary drilled holes were positioned so as to triangulate the site. To investigate smaller scale local variations a 60m grid of 27 No. static cone penetrometer tests was added as an independent check on the uniformity of the site. These were done with a 15cm^2 cone linked to a punched tape recording system which recorded point resistance and sleeve friction at 2cm intervals. This system allowed for the rapid computation of average point resistance over increments of one metre in level.

The non-standard 15cm^2 cone was used because of its ability to penetrate further on this site than the more usual 10cm^2 cone presumably because of reduced friction on the standard diameter rods. The large cone was adopted as a site standard only after preliminary correlatory tests had demonstrated **practical agreement** between the two sizes at several positions and over the maximum possible depth range.

The results of the penetrometer tests are shown in fig. 6. where the (mean - 1 x standard deviation) cone resistance is shown for all tests in the various level increments. The resistance plotted has been reduced by a factor of 4 for comparison with the small diameter borehole SPT results.

Comparison between figs. 5. and 6. and more detailed examination of cone results alongside the three boreholes shows that a correlation factor of 4 is applicable to this site.

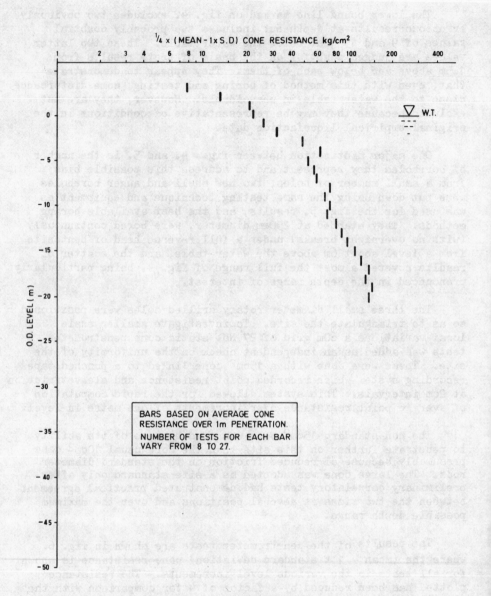

FIG.6 1980 STATIC CONE PENETROMETER RESULTS

Whilst in any 1 metre increment over the depth of interest the
standard deviation of cone resistance varied from 20 to 35% of the
mean, close inspection of the full 2cm frequency plots revealed
no significance in this variation in terms of less dense areas or
levels and it was felt that the static cone tests were a most
efficient method of demonstrating the overall uniformity of a site
for this purpose.

LABORATORY TESTING.

One obvious anomaly in the pre-1980 laboratory test results
received some attention. This was the unsurprising discrepancy
between in-situ relative density implied by the new SPT and static
cone results and those measured previously in the laboratory.

Clearly, one obvious difficulty was the assessment of in-situ
density which in the past had been measured on "undisturbed"
samples as they were withdrawn. To some extent, this was addressed
by borehole geophysical logging and no doubt the probing technique
mentioned by Mr Koning will be of assistance in future studies.
However, one factor we were able to correct fairly simply was the
laboratory-measured minimum porosity.

Conventionally in the U.K. these tests are done by compacting
the material in three layers into a standard mould using a
pneumatic or electric hammer. Sometimes the test is done dry and
sometimes immersed in water. It was felt that such methods applied
to the shelley sands at this site would have produced crushing and
consequentially a modified grading and reduced minimum porosity.
Therefore vibratory densification tests were done on three samples
to re-define the minimum porosity. The tests followed the
procedures described on a number of occasions by Professor
Kolbusjewski and most recently in 1965 (5). They were done
several times for each sample, demonstrating good repeatability
of results, and were followed by a conventional minimum porosity
test. The observations are shown in Table 1 and it was noted from
grading test results that whilst the Uniformity Coefficients before
and after the conventional test remained unchanged in one case,
for the other two samples it increased by approximately 10%.

With the good agreement between the three randomly chosen
samples from widely spaced horizons tentative correction has been
made to the pre-1980 laboratory relative densities by re-defining
minimum porosity as being not lower than 0.35. The effects are
shown on fig. 7. This goes some way towards resolving the
anomalously low laboratory results but the probably larger effect
of low apparent in-situ densities remains.

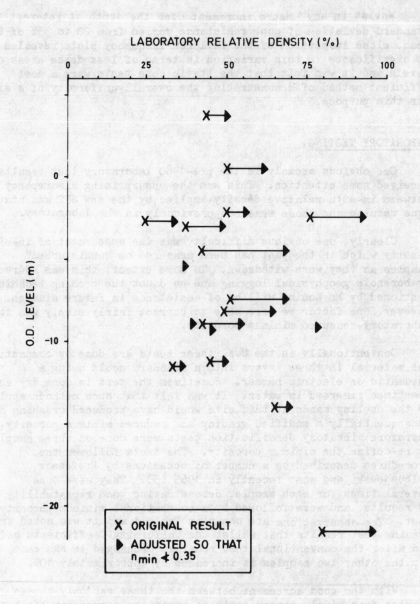

FIG.7 EFFECTS OF REDEFINING MINIMUM POROSITY FROM VIBRATORY TESTS

Table 1 : Minimum Porosity Determinations.

Sample Mean Level	Sample Type	Specific Gravity	Sphericity		Minimum Porosity	
			a*	b**	Vibration Test	Conventional Test
-8.3 m.O.D.	Bishop	2.66	.805	.785	.357	.316
-25.1 m.O.D.	Bishop	2.66	.806	.789	.351	.301
-43.2 m.O.D.	Bishop	2.68	.805	.757	.351	.295

* Sphericity (a) is for plan outline of flat lying discoid particles.
** Sphericity (b) is for estimated section outline of flat lying discoid particles.

CONCLUSIONS.

The following general conclusions may be drawn:

(i) Use of empirical liquefaction potential analyses requires duplication of the boring and testing methods used in the original data sets.

(ii) Static cone penetrometer soundings are an efficient and economic method of demonstrating the validity of an assumption of site uniformity. Clearly, they are a superior technique to the SPT and it is to be hoped that eventually empirical data on liquefaction will permit direct assessments using the static cone.

(iii) Attempts to measure directly the relative density may be in error not only because of the difficulties inherent in estimating the in-situ density, but also because inappropriate laboratory techniques are used to define the limiting porosities.

ACKNOWLEDGEMENT.

The speaker is grateful to the C.E.G.B. for permission to present these results.

REFERENCES.

(1)   PECK, R.B. : 1979, Journal of Geotechnical Engineering
       Division, ASCE, GT3, pp. 393-398.

(2)   WEST, R.G. and NORTON, P.E.P. : 1974, The Icenian Crag of
       South East Suffolk. Phil. Trans. Royal Soc., 269, B895,
       pp. 1-28.

(3)   FUNNELL, B.M. : 1981, Private Communication.

(4)   SEED, H.B. : 1979, Journal of Geotechnical Engineering
       Division, ASCE, GT2, pp. 201-255.

(5)   KOLBUSJEWSKI, J. : 1965, Sand particles and their density.
       Lecture given to the Materials Science Club's Symposium on
       Densification of Particulate Materials. London.

MODELING OF STRONG GROUND MOTION FROM EARTHQUAKES IN THE NORTH SEA AREA

G. Nolet

Vening Meinesz Laboratory
P.O. Box 80.021
3508 TA Utrecht
The Netherlands

Abstract: this paper contains calculations of the horizontal
ground motion that may result from earthquakes in the North Sea
region. Two different models of the North Sea crustal structure
are investigated. The synthetic seismograms contain all wide-
angle reflections, refractions, their multiples and surface
waves in the frequency band between 0 and 1.25 Hz. Simple
approximate formulae, relating signal duration with epicentral
distances, and peak or rms displacement or accelleration with
magnitude, distance and sedimentary cover are presented. Due to
the relatively large depth to the basement, ground motion in the
North Sea is weaker than in the Western U.S. by a factor of
about 3.

1. INTRODUCTION

To determine an acceptable level for the seismic risk of
engineering structures, scaling functions for the prediction
of strong ground motion are an absolute necessity. In land
areas with high seismicity and a dense network of accelerometers, empirical functions may be derived (1). No such
empirical data are available in the North Sea area. By lack of
a better alternative, theoretical seismograms can provide a data
set for the analysis of ground motion. The purpose of this paper
is to calculate such a data set and to derive scaling functions
from these seismograms.

A. R. Ritsema and A. Gürpınar (eds.), Seismicity and Seismic Risk in the Offshore North Sea Area, 303–310.

2. THEORETICAL SEISMOGRAMS

The seismograms in this paper were calculated with the locked mode approximation (2). In this method one sums all the surface wave modes of a horizontally layered model to obtain the time signal. Body waves and leaky modes are forced to enter the domains of discrete modes by introducing a perfect reflector at some depth into the model. The 'approximate' nature of the calculation is merely due to the fact that this reflector gives rise to small reflections at late times that may not be realistic, and that one assumes an upper limit for the horizontal phase velocities for reasons of computational efficiency. Here this limit was taken at 10 km/sec. Physically, this means that reflections leaving the source with an angle smaller than 20° were excluded. In practice it puts a lower limit of about 10 km to the epicentral distance that can be considered. I have limited the frequencies to 1.25 Hz, so that eigenperiods of oil platforms fall well within the frequency interval.

3. NORTH SEA MODELS

For the calculations I used two models. Model A is an averaged model for the North Sea, derived from surface waves by Stuart (3). This model has a strong resemblance to the structure found off the coast of England with refraction studies (4). Model B is a slightly simplified version of the Doggerbank crustal structure that was interpreted on the basis of reflection data (4,5). The models differ mainly in the thickness of the sedimentary layer and the compactness of the sediments near the surface. Anelastic damping was introduced by assigning a quality factor Q to each layer.

	depth(km)	v_s(km/sec)	density(g/ccm)	Q
model A	0 - 5	2.300	2.40	50
	5 - 13	3.500	2.70	500
	13 - 30	3.875	2.80	500
	> 30	4.650	3.25	200
model B	0 - 1.8	0.840	2.05	20
	1.8 - 3	2.590	2.20	50
	3 - 7.4	3.030	2.30	100
	7.4 - 24.4	3.730	2.70	500
	24.4 - 30	3.940	2.80	500
	> 30	4.650	3.25	200

4. SOURCE MODEL

The source was placed at the top of the first layer where the S-velocity exceeds 3.25 km/sec (v_p > 5.6 km/sec.), i.e. at a depth of 5 km in model A, and 7.4 km in model B. The high water content and the low strength of the sediments in the upper layers make it unlikely that stresses can accumulate to appreciable levels in these sedimentary layers (6), and note-worthy earthquake activity will be confined to the basement, as in the Gulf of Mexico (7). A fault with a dip angle of 60°, and a strike-slip source was used (in this geometry, horizontal peak accelerations for dipslip are less than half of those found for strike-slip sources). I used a kinematic fault model. In the present study there is no need for more sophisticated dynamic models, such as the one recently proposed by Vlaar (8). The model is a simple Haskell fault model with an unidirectional propagation of rupture; in this model the spectrum of the slip function in the direction of maximum radiation is:

$$F(\omega) = M_o \omega_c \frac{1-\exp(i\omega/\omega_c)}{\omega^2(i\omega/\omega_c-1)} \qquad (4.1)$$

where M_o is the seismic moment and ω_c the corner frequency. This model satisfies the ω^{-2} dependence for body waves, which is generally observed (9) in the far field. Since accelerations are highest for high frequency, we expect peak accelerations to be found near the corner frequency, or, if ω_c is too large, near the upper end of the frequency interval (1.25 Hz). Thus the choice of ω_c introduces a nonlinear behaviour of peak accelera-tion as a function of M_o as soon as ω_c enters our interval of interest. Madariaga (10) finds for his fault model:

$$\omega_c = 0.21 \, v_s/L \qquad (4.2)$$

where L is the source radius. If we adopt the following relation between L and the stress drop s (11):

$$s = \frac{7}{16} M_o L^{-3} \qquad (4.3)$$

and assume s to be 10^7 Pa (100 bars) we find the following approximate relationship between ω_c and M_o :

$$\omega_c = k \, M_o^{-1/3} \quad rad/sec \qquad (4.4)$$

where $k = 2.1 \times 10^5$ if M_o is in Nm.

Since it is more practical to work with magnitudes than with seismic moment, I shall in the following 'translate' seismic moments to local magnitudes M_L, using a scaling law found for intraplate earthquakes in Germany (Ahorner, pers.comm. 1982):

$$\log M_o = 10.4 + 1.1 \, M_L \qquad Nm \qquad\qquad (4.5)$$

Historically, strong events with magnitude up to 6 have been reported. Accurate magnitude determinations (by the ISC) are only available from 1964 onward. A search among offshore events between 5^oW and 10^oE in the ISC bulletins between 1964 and 1979 showed a maximum magnitude of 4.9 (m_b ; this event was on July 22, 1967 at $51^o.4$N, $1^o.3$E). In this paper I shall study events with magnitudes up to 6.

5. RESULTS

I have calculated about 100 theoretical seismograms for epicentral distances $10<R<150$ km, magnitudes $4.2<M_L<6.0$ and source depths $5<H<30$ km. Figure 1 shows 3 representative examples for $M_L=6$. The first two signals are at the same distance of 150 km, but the source depth is 5 km in (1a) and 30 km in (1b). At this distance the surface waves have had sufficient time to develop, and a wavetrain with a length of about 16 seconds is visible in (1a). The dominant period is between 2 and 5 seconds, coinciding with the fundamental eigenperiod of some rigid oil platforms. However, the accelerations involved are small - of the order of 0.002g. A deeper source (1b) does not excite the surface waves. An important observation is that the longer period surface wavetrains are not yet developed near the source (1c); here the large peak, arriving at 15 seconds, has a dominant period of 1.3 seconds.

From these seismograms I have derived a scaling law to estimate the highest values to be expected for accelerations and displacements. From 70 seismograms with source depth at 5 km (model A) and 7.4 km (model B) I determined the peak values of acceleration and displacement, and the root-mean-square (rms) average over the first 100 seconds.

Figure 1. Synthetic seismograms calculated for model A, for a frequency band of 0-1.25 Hz, a seismic moment of 10^{17} Nm, and several values of the source depth H and epicentral distance R. (a)-H=5 km, R=150 km, (b)-H=30 km, R=150 km, (c)-H=5 km, R=30 km.

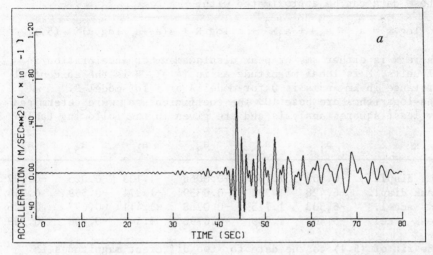

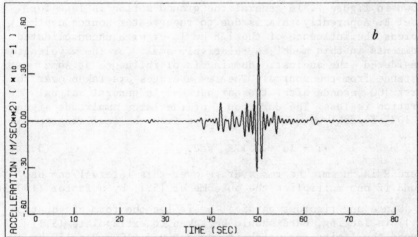

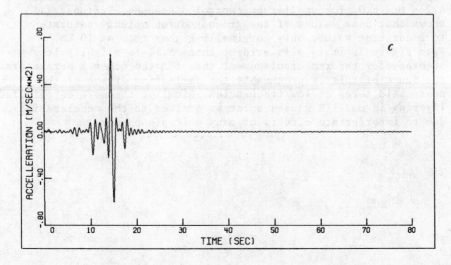

These data can be approximated with:

$$\log x = a_1 + a_2 M + a_3 M^2 + a_4 \log R + s(a_5 + a_6 \log R) \qquad (5.1)$$

where x is either rms or peak displacement or acceleration in
SI units, M is local magnitude as in (4.5), R is the epicentral
distance in km and s is 0 for model A or 1 for model B.
The logarithms are base 10. The coefficients a_j were determined
by least squares analysis and are given in the following table:

x	a_1	a_2	a_3	a_4	a_5	a_6
rms displ.	−6.963	1.155	−0.0066	−1.135	0.025	−0.097
peak displ.	−6.428	1.392	−0.0339	−1.274	−0.569	0.214
rms accel.	−6.503	1.715	−0.0788	−1.411	−0.697	0.226
peak accel.	−5.409	1.739	−0.0798	−1.547	−0.936	0.369

The fit of (5.1) to the data for two different magnitudes is
shown in figure 2. In general the ground motion is less for
model B. Apparently this is due to the greater source depth,
whereas the influence of the 1.8 km layer with unconsolidated
sediments in this model is relatively small for the wavelengths
considered (the anelastic damping is of influence at some
distance from the source). The rms averages are taken over the
first 100 seconds after the earthquake. In general, signal
duration is less. The duration T of the large amplitude signal
(> 10% of peak) was estimated by eye from the accelerograms:

$$T \approx 10 + 0.17R \ \text{sec.} \qquad (5.2)$$

where R is in km. The rms average over this interval can be
found if one multiplies the outcome of (5.1) by a factor $(100/T)^{\frac{1}{2}}$.

Since any rigorous physical basis for the form of equation
(5.1) is lacking, care should be taken to extrapolate (5.1)
beyond the limits for which it was derived: 10 < R < 150 km,
4.2 < M < 6.0. For smaller epicentral distances, Trifunac (1)
shows that peak values of the ground motion quickly saturate
to a constant value, only marginally higher than at 10 km.
From figure 2, it is also evident that (5.1) is slightly less
accurate for the rms displacement than for the other 3 parameters.
The cause of this is presumably the dominating effect of the
near-field terms at low frequencies. Finally, the more complicated
layering in model B causes a larger scatter in the accelerations,
due to interference effects of many multiple phases.

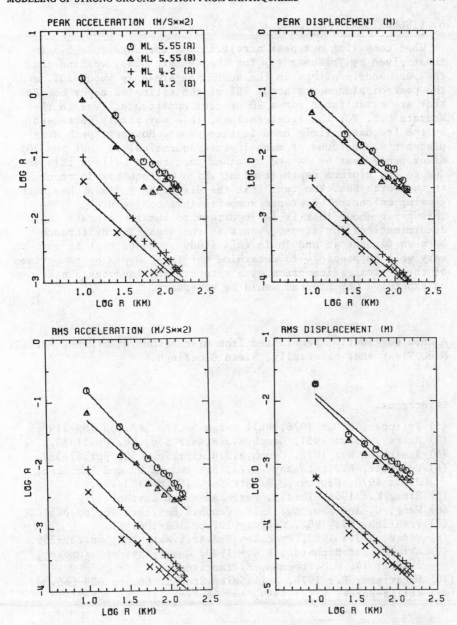

Figure 2. Fit of the data for M = 4.2 and 5.55 and Models A
and B to equation (5.1), given as solid lines.

6. DISCUSSION

When comparing our peak accelerations for magnitude 5.5 to
those given by Trifunac (1) for the Western U.S., we find that
the peak accelerations in the North Sea are only about 20%, and
the peak displacements about 30% of the estimated upper bounds
that are established for a 90 percent confidence level in the
Western U.S. For the accelerations, this may partly be caused
by the frequency limit used in this paper. However, peak displacements are found at much lower frequencies, so that the 30%
discrepancy must be caused by other factors. I believe this is
due to the minimum depth level at which earthquakes do occur
in the North Sea. The fact that the discrepancy diminishes with
growing epicentral distance supports this conjecture.
This paper shows that it is important to obtain accurate depth
determinations for strong events in the area. The difference
between figures 1a and 1b in this study suggests that it may
very well be possible to determine the focal depth by comparison
of theoretical seismograms with records from land-based stations;
no expensive OBS program would be necessary.

Acknowledgement: I benefitted from discussions with Rinus Wortel,
Nico Vlaar and, especially, Sierd Cloetingh.

References.

(1) Trifunac, M.D.: 1976, Bull.Seism.Soc.Am. 66, pp.189-219.
(2) Harvey, D.J.: 1981, Geophys.J.R.astr.Soc. 66, pp.37-70.
(3) Stuart, G.W.: 1978, Geophys.J.R.astr.Soc. 52, pp.367-382.
(4) Collette, B.J., Lagaay, R.A., Ritsema, A.R. and Schouten,
     J.A.: 1970, Geophys.J.R.astr.Soc. 19, pp.183-199.
(5) Hinz, K.: 1964. Thesis, Bergakademie Clausthal.
(6) Wang, C. and Mao, N.: 1979, Geophys.Res.Lett. 6, pp.825-828.
(7) Frohlich, C.: 1982, Geology 10, pp.103-106.
(8) Vlaar, N.J.: 1982, Proc.Kon.Ned.Ac.v.Wet. B84, pp.239-256.
(9) Aki, K. and Richards, P.G.: 1980, *Quantitative Seismology*,
     chapter 14, W.H.Freeman, S.Francisco.
(10) Madariaga, R.: 1976, Bull.Seism.Soc.Am. 66, pp.639-666.
(11) Keilis-Borok, V.I.: 1959, Am. Geofis. 12, pp. 205-214.

DISCUSSION-NOTES SECTION 5

Soilmechanics, Liquefaction, Geotechnology

Q. D.N. Cathie:

Since vertical accelerations may be significant for the
design of certain parts of the structure we should consider
whether a sealed horizontal spectrum is adequate and also
consider carefully the peak vertical acceleration.
We have heard from Dr. Gudmestad of the giant fixed platforms under consideration which will have longer natural
periods of vibration than earlier platforms. However, it
is probable that modes of vibration excited by vertical
motion will still be those corresponding to quite high
frequencies. Therefore, the relatively high frequency content of the vertical time history will be particularly
important alongside the lower frequency component of the
horizontal time history.
The level of the vertical peak ground acceleration and the
response spectrum used will also be relevant to the bearing
capacity analysis for the foundation of a gravity structure.

A. P.B. Selnes:

I fully agree with your points. The vertical component has
received little attention in the past, because it is
generally of minor concern in onshore design. This is not
satisfactory for offshore design.

Q. G. Woo:

Please, could you discuss further the importance of vertical motion for offshore design, especially cantilevered
structures.

A. O.T. Gudmestad:

According to the API rules "an acceleration spectrum of
one - half that for the given zone should be applied in
the vertical direction. All three spectra should be applied
simultaneously".
For the vertical component of the earthquake at the mudline
the interaction with the seawater has to be taken into
account. For both the vertical and the horizontal components
the modification of the earthquake parameters due to the
seabed conditions must be evaluated as discussed in my
paper.
The modification of the earthquake properties due to the

311

A. R. Ritsema and A. Gürpınar (eds.), Seismicity and Seismic Risk in the Offshore North Sea Area, 311–314.
Copyright © 1983 by D. Reidel Publishing Company.

presence of the structure shall further be accounted for
and the components of the earthquake waves can be applied
in the two horizontal and the vertical direction at the
mudline.
The characteristic response quantities resulting from each
component can be combined using the "square root of sum of
squares" rule.
It is thus considered that the resulting characteristic
quantities consistently account for the vertical acceleration of the earthquake.
The vertical acceleration at the deck level and also in
other parts of the structure might be dominated by the
vertical component of the earthquake. The vertical acceleration might also be higher than the horizontal.
At this time, however, A/S Norske Shell's studies related
to earthquake design are in a very preliminary stage and
we cannot give a quantitative assessment of the effects
of the different components of the earthquake excitation.
I do hope, however, that I later will have the opportunity
to release more information on this matter.

Q. P.W. Burton:
It seems that magnitudes of about $5\frac{1}{2}$-6 occur offshore
Norway. What peak ground motion accelerations do these
models of Dr. Nolet associate with these earthquakes, and
with what uncertainties?

A. A.M.H. Nolet:
If a magnitude 6 would occur, my calculations give a peak
of 0.4g at an epicentral distance of 10 km using Stuart's
model and about half that value in the 'Doggersbank' model.
The question of the uncertainties is important. Trifunac
uses statistics on empirical data to establish confidence
levels. My approach is a deterministic one, and the uncertainties are directly related to the assumptions. Where
factual knowledge was absent, I have maximized the accelerations. For example, the lower crust has very little
attenuation, and the corner frequency was adopted from
Madariaga's earthquake model, which leads to higher acceleration than, for instance, Aki's model. There is one
factor which is difficult to evaluate, and that is lateral
heterogeneity. My opinion is that it will have small effects
close to the source since one needs to be several wavelengths away from focussing effects to become significant.
Channeling effects in the Viking graben can only develop
after distances large compared with the width of the central
part of the graben, which is about 30 or 40 km. Considering
the relatively weak magnitudes in the area, the uncertainty
becomes less relevant at those distances. Altogether I

think my calculations give maximized ground motion to a
very high confidence level, maybe even too high.

Q. Y. Zaczek:
For that study of nuclear power plant, Standard Penetration
Tests were performed down to 60 meters. Do you use the
hammer at the top or at the bottom of the boring?
For the SPT, with the hammer at the top, performed at
great depth you measure the elasticity of the rods and
not the soil properties.
TERZAGHI recommends to perform such SPT tests in the top
layers, let's say about 20 meters depth.

A. D.J. Mallard:
I must stress that we are using empirical relationships
and therefore had to attempt to duplicate the original
tests which contributed to the liquefying/non-liquefying
criteria. These did not include down-hole SPT hammers.
Where precise details of the original tests were un-
available we attempted always to choose a conservative
option.
With regard to SPT methods as such, whilst I would agree
that energy losses in the rods become an increasingly
significant factor with depth, I am not sure that with our
rods this would be of significance in the top 20 m. Also,
whilst there are many shortcomings associated with SPT
hammers at the surface, I tend to feel that use of a down-
hole hammer at depth under water introduces an even larger
degree of non-standardisation making the use of any empirical correlation doubtful.

Q. A.R. Ritsema:
In relation to the long-period characteristics of new
structures in the North Sea basin and an eventual repetition
of the Lisboa earthquake of 1755, it seems important to
study long distance effects of very large earthquakes at
long-period structures, such as existing multi-store
buildings. Are records of this kind known, f.e. from the
latest large magnitude earthquake, that of Sumbawa Island
in 1977, August 19, felt at distances of more than 2000 km?

A. N.N. Ambraseys:
From this particular event no records do exist, but cracks
developed in high apartment-buildings in Perth, W-Australia
and liquefaction occurred in the harbour dam.

Q. A. Gürpinar:

In considering the possible benefit one may expect from
allowing plastic soil deformation to reduce accelerations
in the structure, what is the degree of control the
engineer has in controling this deformation at an acceptable level?

A. P.B. Selnes:

To be able to establish upper bounds of movements which
may take place is a prerequisite for utilizing design
concepts allowing plastic yielding in the foundation.
What we know at present is there may be considerable
economic benefits in such a design, it is therefore
worthwhile to pursue further research on this topic. Such
research may involve both analytical procedures and dynamic
model testing. And, already today, I feel that we have
analytical tools that may allow some plastic yielding
(irrecoverable displacements) to take place provided the
margin of safety is sufficiently high.

Q. A. Gürpinar:

Regarding the problem of soil amplification, should this
be taken as an amplification of the peak ground acceleration
as suggested by Koning for example, or a spectral amplification as suggested by API influencing only the period
spectral ordinates?

A. H.L. Koning:

In what way soil amplification is expressed depends on the
problem tackled. The writer used the amplification of the
peak ground acceleration merely as a simple expedient to
compare some accelerograms. In estimating the seismic risk
the complete record will be employed.

Section 6

RISK ANALYSIS

EVALUATION OF SEISMIC RISK

N.N. Ambraseys

Imperial College of Science and Technology
London, England

1.0.0. INTRODUCTION

The last 15 years have seen a very rapid increase in the
demand for earthquake resistant design of structures not only in
the more seismic regions but also in the less earthquake prone
areas of northwest Europe, where the engineering profession is
being involved in an ever wider range of project types that require
earthquake resistant design or construction. An increasing number
of professionals are also becoming conscious of earthquake risk
and require risk assessments made for them, even in low seismic
hazard regions, such as the North Sea and surrounding countries
(Belgium, Denmark, Germany, The Netherlands, Norway, and the
United Kingdom) for such petrochemical installations as LNG and
LPG storage, chemical processing plants, off-shore structures and
nuclear power plants.

The general consensus of opinion among engineers in north-
west Europe is that without an awareness and useful knowledge of
engineering seismology and earthquake engineering, the engineer
abroad, and to a lesser extent at home, is at a disadvantage.
With an ever increasing competition in the international market,
engineering seismology is considered to be an invaluable asset
and at the same time a substantial invisible export.

The rapid development of important engineering structures
and increasing demand for the solution of complex structural
problems is already beginning to cause concern. One of the causes
for concern is the lack of reliable data for the selection of the
appropriate design parameters. Considering that industrial developments are spreading into areas of little-known seismicity such as

317

A. R. Ritsema and A. Gürpınar (eds.), Seismicity and Seismic Risk in the Offshore North Sea Area, 317–345.
Copyright © 1983 by D. Reidel Publishing Company.

the North Sea and in particular that time rarely allows for the
acquisition of earthquake data or the re-evaluation of existing
seismic information, the engineer is likely to be forced, on
occasions, to step across the hazy borderline of safety by
accepting an element of risk over and above what he would otherwise have considered to be normal.

For ordinary structures of relatively short life this risk
may, under certain conditions, prove to be economically acceptable.
However, for special structures whose failure or damage during an
earthquake may lead to uncontrollable disasters, this attitude is
not acceptable. To determine what is an acceptable risk for a
particular engineering structure, a considerable amount of
informed judgement, detailed technical evaluation of the structures
and great caution must be employed in the selection of earthquake
design parameters. The assessment of hazard and the selection of
the design ground motions or dynamic loads, appropriate to a
given geographical region and local foundation conditions, is
considered to be one of the outstanding problems.

The decision to undertake measures to protect engineering
structures from earthquake damage is usually based on assessments
of the risks to the community and to its economy as well as on
judgement as to whether these risks are acceptable. The data
needed for the quantification of risks is at present insufficient
to support risk predictions with a high degree of certainty.
However, we do have a conceptual basis for progress in this field.

2.0.0. EARTHQUAKE RISK

Earthquake risk may be defined as the probability of the
loss of property or loss of function of engineering structures,
life, utilities, etc. The factors entering into the assessment
or quantitative estimation of earthquake risk are, the Seismic
Hazard, i.e. the probability of occurrence of groundmotions due
to an earthquake, the Value of the elements exposed to the hazard,
i.e. lives and property, and the Vulnerability of these elements
to damage or destruction by the ground motions associated with
the Hazard.

In their simplest functional form the factors entering into
the assessment of risk may be expressed by the following
relation:

(Risk) = (Seismic Hazard)*(Vulnerability) (Value)   (1)

This relation may be applied to estimate the risk in financial
or economic terms, to any object or set of objects, i.e. engineering
structures and their contents, a utility network, as well as to
assemblies of such objects.

2.1.0. Seismic Hazard

Seismic Hazard is the probability of occurrence, at a given place or within a given area and within a given period of time, of ground motion due to an earthquake of a particular magnitude or size capable of causing significant loss of value through damage or destruction. Seismic Hazard may normally be expressed analytically or by sets of curves depicting the probability of occurrence of certain ground accelerations, velocities or displacements, of ground movements of various durations, or any other physical parameter (x) which is significant for assessing the vulnerability of a structure. The distribution of (x) is often expressed in terms of the probability H(x) or (x) being exceeded within a given period of time, or in terms of h(x) = 1 - H(x), which is the probability of (x) not being exceeded or in terms of its probability density function

$$h'(x) = \frac{dh}{dx} .$$

Seismic Hazard is beyond human control, but some knowledge of it, as well as of its spatial distribution and fluctuation in time, is possible through the seismo-tectonic study of the region and the evaluation of its long-term seismicity. The assessment of hazard is the subject-matter of tectonics and seismology as well as of engineering seismology which is the link between earth sciences and engineering.

The problem of making estimates of Hazard, that is of the probability of occurrence, within a given period of time (usually determined by the useful life of the structure in question at a given site or within a given area) of earthquake ground motions capable of causing significant damage, involves:
(a) estimating the probability of occurrence of earthquakes of various magnitudes at various focal distances from the site and their maximum value;
(b) identifying the probable location of focal regions and mechanism of the events;
(c) determining the attenuation factors affecting the propagation of seismic waves between the probable earthquake source area and the site; and
(d) estimating the influence of the local foundation conditions and of the structures themselves on the characteristics of strong ground motion at the site.

2.1.1.0. (a) Estimation of the probability of occurrence of earthquakes of various magnitudes at various distances from a site requires the knowledge of (i) the earthquake foci, and (ii) their size and mechanism.

2.1.1.1. (i) Proximity of an earthquake source to a particular engineering site is usually measured from the focus of the event which is defined by its epicentre and focal depth.

Instrumental recording of earthquake began very late in the last century and continued for some time with instruments which were very imperfect. Attempts to locate earthquake epicentres instrumentally also date back from the turn of the century, with results which are very unsatisfactory. The determination of epicentres does improve after the early 1930s but it remains relatively poor up to the late 1950s, and locations estimated by the International Seismological Summary (ISS) and the Bureau Central International Seismologique at Strasbourg (BCIS) are on the whole of low accuracy. Between 1920 and 1950 the accuracy of location of the larger shocks does improve, but for the smaller shocks location errors remain large. This is partly due to the fact that during the period 1918 to 1953 about 60% of the ISS determinations per annum were "adopted", i.e. not calculated. It is only after the mid-1960s that the location of epicentres becomes reasonably accurate.

Epicentres of early events may be relocated using either routine computer methods or more sophisticated techniques. However, the success of the relocation will depend not so much on the method used as on the quality of the input data. For earthquakes before the mid-1920s the input data is so poor that relocation is very difficult. After that time and for the rest of the first half of this century, routine relocations may improve the overall accuracy. Relocations determined by either Joint Epicentre Determination (JED) techniques, applied to geographically limited areas, or singly with travel-times modified by JED-calculated source-station adjustments in which macroseismic data is used to reduce bias, may show a significant improvement in epicentral estimates, Figure 1.

A reliable method for improving or confirming instrumentally determined epicentres on land of pre-1960 earthquakes is to attempt to find correlations with macroseismic reports. The true source of the earthquake may not be in the centre of the meizoseismal area, and the meizoseismal area is very dependent on population distribution; but at least the macroseismic epicentres will not be liable to the gross mislocations which are possible with instrumental locations.

The determinations of reliable focal depths on a routine programme, particularly for events prior to about 1963 is difficult. This is because early arrival data are generally of such poor accuracy and azimuthal distribution that the resolution of focal depth is likely to be even worse than that of the epicentral locations.

Thus, for the study of regional tectonics and siting of
engineering structures, an indiscriminate use of instrumental
locations of epicentres over the period 1906-1979 is not recommended. For the engineer, proximity to a site of the epicentre
of an earthquake located on land instrumentally is a cause for
concern only when the accuracy of its location is high and if
possible, when this is confirmed by macroseismic evidence.

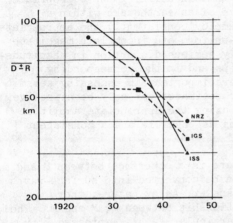

Figure 1. Average distance D
between macroseismic epicentre
and instrumental epicentres in
Iran located by ISS, IGS and
Nowroozi. (M_s > 5.0)

(see: Geoph. J.R. Astr. Soc.,
53: 117-21, 1978)

2.1.1.2. (ii) Earthquake magnitude is usually considered to be
the parameter that measures the amount of energy that radiates
away from the focal volume of an earthquake, a measure of the
size of the event that can be assessed by several different scales.
For instance, the body-wave magnitude (m) is based on the amplitude
of waves with periods of around 1 second. The local magnitude (M_L)
as defined by Richter is based on the maximum trace amplitude
of the standard Wood-Anderson seismograph. Consequently, different
magnitude scales assess the rate at which seismic energy is
released within different frequency bands, rather than the total
earthquake energy release. One needs to know more about the
seismic spectrum to be able to assess the total energy release
and care must be taken to identify the scale to which a magnitude
refers.

It was only after the early 1950s that the concept of magnitude began to receive general attention, so that earthquakes
before that period were not assigned magnitude values. During the
next decade, seismological stations began to make their own
magnitude determinations from different wave types so that by the
mid-1960s there was a multitude of formulae for magnitudes. Starting with 1962 the USSR network standardised its magnitude and
since then M-values are reported on a routine basis. A year
later, the International Seismological Centre (IGS, the successor
of the ISS) and the USGS began to publish m-values regularly,

while many local stations continued to calculate M_L estimates
which are quite numerous but grossly heterogeneous. Thus, for the
whole period up to the mid-1960s we have an agglomeration of
magnitudes estimated at different times by different scales,
often by a combination of scales, or by scales that are not
specified.

It is important therefore, that in defining the appropriate
parameter that describes the size of an earthquake within a
particular frequency range, we employ consistently the appropriate
magnitude scale. The use of non-uniform and incomplete magnitude
estimates may cause a potentially serious distortion in magnitude-
frequency recurrence relationships and in the corresponding
expressions for $h'(x = M)$ that are important in assessing seismic
hazard. Unification of magnitude estimates can only be effective
if magnitudes are re-calculated using the appropriate amplitude-
period data from as many stations as possible and a consistent
method of calculation.

The question of whether there is a coherence between M and m,
so that body-wave magnitudes can be converted into surface-waves
and vice versa is difficult to answer. As by nature, there
should be no general valid relationship between m, M and M_L. The
high-frequency radiation of a particular event, expressed by
(m) cannot be uniquely related to the low frequency radiation of
the same event given by (M) without introducing additional
considerations about the material properties and mechanism at
the source, including stopping phases, barrier intervals or
source multiplicity. In other words, a seismic source may produce
a larger short-period radiation than long-period, and this will
be reflected in the values of (m) and (M), while another source
may show the reverse repartition of radiation. In the first
instance we may consider, as in optics, that the source is near
the "blue" end of the spectrum, and in the latter, near the "red",
earthquakes ranging from predominantly "hard" or "blue" to predominantly "soft" or "red".

Unfortunately, neither M, m nor M_L give an absolute measure
of the size of an earthquake. Seismic energy, being the result
of a transient energy release in the earth's crust through slip
or dislocation occurred and on the average degree of dislocation.
These controlling factors are in fact combined in the seismic
moment of the earthquake, M_O, which is a better measure of the
size of an earthquake but which can be calculated for recent events
only, of the last 10 years, and for a few large earlier events.

2.1.2.0. (b) The identification of probable locations of future
earthquakes and their spatial distributions requires knowledge
of the local tectonics. Almost all very shallow earthquakes of
magnitude in excess of about M = 6.0, on close examination, have

been found to be associated with surface faulting.

Surface ruptures are neither continuous along the whole
length of a fault-break, nor do they follow precisely mapped
faults. They seem rather to follow a path of least resistance
with a comparatively broad shear zone, from a few hundreds of
metres to a few kilometres wide, shifting laterally from one
shear plane of weakness in one part of the zone to another elsewhere, in most cases with well developed large-scale en echelon
patterns, tension features and grabens are connected with en
echelon shears.

These details will show if the fault zone is mapped on a
large scale. Mapping on a smaller scale would tend to obliterate
the details of the actual shear pattern and the fault would appear
a continuous and smooth trace. Widths of shear zones associated
with strike-slip faulting range up to 2 kilometres, while widths
in normal faulting may be twice as broad, up to 5 kilometres.
For thrust, shear and fracture zones can be much wider, depending
on the dip of the fault plane.

Displacements, both horizontal and vertical on individual
ruptures within the fault zone can be large, with the formation
of grabens and pressure ridges with fault zones. Usually the
width of a fault zone is narrowest where strike-slip fault-breaks
occur, becoming wider for normal and thrust fault-breaks. Fault
ruptures and ground deformations associated with a fault zone
may have serious consequences for engineering structures situated
within these zones.

Within active fault zones, creep may cause additional damage
to structures before or after an earthquake. Fault creep or
slippage within fault zones is known to have occurred aseismically
in a number of active zones with post-earthquake movements and
with aseismic slippage at a rate of about 1 cm/year.

Using geological, seismological and historical data is it
often possible to assess the relative activity of a geological
fault, and classify it in one of the following categories: (a)
active, (b) potentially active, (c) uncertain activity, and (d)
inactive. The following system for classification of fault
activity may be used:

2.1.2.1. (a) Active faults: historical or recent surface faulting
with associated strong earthquake; tectonic fault creep or geo-
detic indications of fault movement; geologically young deposits
displaced or cut by faulting; fresh geomorphic features characteristic of active fault zones present along the fault trace;
physical ground water barriers in geologically young deposits;
stratigraphic displacement of Quaternary deposits by faulting;

offset streams. Seismological earthquake epicentres are associated
with individual faults with a high degree of confidence, including
historical earthquakes that were felt over a large area.

2.1.2.2. (b) Potentially active faults; no reliable report of
historic surface faulting; faults are not known to cut or displace
the most recent alluvial deposits, but may be found in older
alluvial deposits; geomorphic features characteristic of active
fault zones are subdued, eroded and discontinuous; water barriers
may be present in older materials; the geological setting in
which the geometric relationship to active or potentially active
faults suggest similar levels of activity. Seismically there is
alignment of some earthquake foci along the fault trace, but
locations are assigned with a low degree of confidence.

2.1.2.3. (c) Faults of uncertain activity; available information
is insufficient to provide criteria or definitive enough to
establish fault activity.

2.1.2.4. (d) Inactive faults; no historical activity based on a
thorough study of local sources of information. Geologically,
features characteristic of active fault-zones are not present,
and geological evidence is available to indicate that the fault
has not moved in the present past. Seismologically, not recognised
as a source of earthquakes.

Criteria for recognising an active fault may be summarised
as follows:
Geological: An active fault is indicated by young geomorphic
features such as fault scarps, fault rifts, pressure ridges, offset streams, enclosed depression, fault valleys, and ground
features such as open fissures, "mole tracks", rejuvenated streams,
folding or warping of young deposits, ground water barriers in
recent alluvium, en echelon faults and fault paths on recent
surfaces. Usually a combination of these geomorphic features is
generated by fault movements at the surface. Erosion features
are not necessarily indicative of active faults, but may be
associated with some active zones.
Seismological: Large magnitude earthquakes and microearthquakes, when fairly precisely located or relocated instrumentally,
authenticated macro-seismically, may indicate an active fault.
Earthquake swarms associated with volcanic activity are not
necessarily indicative of active faulting. A lack of known earthquakes, however, should not be used to indicate that a fault is
inactive.
Historical: Historical sources examined at first hand. Fault
movements or creep may be detected from displaced man-made
lineaments such as early canals, surface or underground, roads,
archaeological sites, railway and power lines.

Faults are considered to be significant to a site if they cross it or if they are near-by and could generate potentially damaging earthquakes.

The mobility of a fault may be assessed in terms of local tectonics. For the Middle East, the following relation has been found between surface-wave magnitude of the event (M_s), the associated length of rupture (L) and maximum surface displacement (R) in centimetres:

$$M_s = 1.1 + 0.4\log(L^{1.58}R^2) \quad \text{for } 7.4 \geqslant M_s \geqslant 5.0$$

mainly from strike-slip fault-breaks. Maximum displacements (R) have been found to be 3.8 ($\pm$ 0.9) times larger than average displacements along the length of the rupture. This equation may be used in comparable tectonic regions to estimate the expected magnitude which might result from faults of known or inferred length and mobility.

2.1.3.0. (c) Determination of the appropriate attenuation laws affecting the propagation of seismic waves between the source and site is usually based on strong-motion recordings. There are at present more than two dozen published correlations or attempts to derive attenuation laws between ground motion parameters (acceleration or velocity) and focal distance (R), magnitude (M) and site conditions.

All attenuation laws, however, are hampered by the limited amount of strong-motion data, particularly at short distances and for large magnitude events. For example, there is very little data in the near-field (R < 10km) of medium to large magnitude earthquakes (m > 6.0) and no data at all from large magnitude events (M > 7.1). But what is equally serious is that even for the data available the magnitudes or measures of the size of the event used in the derivation of these laws are not homogeneous.

Non-homogeneity of the focal distance (R), particularly in the near-field, is also a serious problem. R-values determined from instrumentally calculated foci are likely to be in error by 5 to 15 kilometres or more, particularly for the smaller magnitude events in regions where the seismic network is not specially designed for high accuracy locations.

Since 1965 a number of important strong-motion records have been obtained from earthquakes in Central and Southern Europe and in other parts of the Middle East but no such records are available for Northern Europe and other areas of the World of comparable seismicity and tectonics.

Nevertheless, what the total body of strong-motion data to

date suggests is that within or near the focal volume of an
earthquake, peak accelerations seem to be independent of magnitude and variations in acceleration within this volume depend on
the mechanism and properties of the material rather than on
distance from the focus. The available data also suggests that
at any range of magnitude the tendency of the maximum peak acceleration is to increase with increasing number of observations.
This tendency is becoming more clear with time, as more observations are becoming available from denser strong-motion networks
and from networks deployed to monitor aftershocks in meizoseismal
areas. Such networks stand a better chance of sampling small
magnitude events whose focal volume is small.

     It is very likely, therefore, that on bedrock sites within
the focal volume, accelerations may reach or exceed 100%g but at
soil sites they might be less because of the limited strength of
the surface deposits, see figure 2. On the other hand, on bedrock
sites, large peak accelerations may be associated with high
frequencies and short duration so that from the engineering point
of view maximum peak acceleration alone is not always the appropriate parameter for design or for scaling design ground motions.

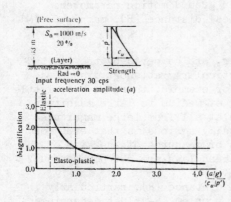

Figure 2. Magnification of bedrock acceleration (a) in a deposit of normally consolidated
material as a function of the
dimentionless strength parameter $(a/g)/(c_u/p')$. Note that
as the undrained strength of the
material c_u decrease or as the
input ground acceleration increases, the response becomes
elastoplastic and the magnification falls off to values of
less than 1.0.
(see: Proc.3rd Europ.Symp.Earthq.
Eng.,$\underline{1}$:309, Sofia, 1970)

2.1.4.0. (d) The assessment of the influence of local soil
conditions and of the soil-structure interaction on the characteristics of ground motions is a problem that requires knowledge not
only of the general earthquake characteristics mentioned above,
but also of the characteristics of the foundation materials well
beyond their elastic range of response.

     The question of whether there is an upper bound for ground
accelerations and velocities is indeed of great importance to the
engineer. It cannot be answered, however, by treating the ground

as a purely elastic, hysteritic medium. This is because the
near-surface foundation materials, through which acceleration
pulses have to travel before reaching the surface, have a finite
shear strength, and stress waves under certain conditions will be
prevented from reaching the surface without substantial reduction.
For competent bedrock within the earthquake source, analysis
predicts maximum accelerations which, depending on the source
parameters, may reach values well in excess of 100%g.

The answer to this question, however, lies in observing the
effects of strong earthquakes on the ground itself. For instance,
non-tectonic cracking of soil and weathered rock, slumping and
settlements, induced by shaking, no matter how small, suggest an
inelastic mechanism of overstressing and failure of the ground
brought about by acceleration pulses capable of producing these
effects; it is precisely this mechanism that in the near-field
distorts recorded high-frequency ground motions to the extent
that they cannot be correlated with purely source-path parameters.

For instance, it can be shown that the maximum possible
acceleration in homogeneous saturated soil deposits cannot exceed
values of the order of about $0.5(c_u/p')$, where (c_u) is the undrained
shear strength of the deposit at an effective consolidation pressure (p'). Thus, an upper bound for the maximum acceleration in
near-surface deposits, as it should, is dictated solely by the
shear strength properties of the material.

For strong foundation materials or for small accelerations
the deposit will behave elastically and it may therefore amplify
the bedrock motion. As the strength of the foundation material
decreases, or as the intensity of motion increased the response
may bring about internal or near-surface yielding, as a result
of which accelerations above a certain level will be prevented
from reaching the surface, and permanent displacements will occur,
Figure 3.

Duration of strong ground motion is one of the most important
factors in producing damage. Prolonged cyclic loading of soils,
even at relatively low stress levels, may cause excessive pore
pressure to develop in saturated foundation materials or progressive
loss of cohesive strength in building materials.

Also, increasingly difficult questions in soil dynamics need
to be answered, e.g. foundation compliance, earth slope stability
and liquefaction potential. Here there is a great imbalance of
knowledge. Methods of analysis are highly developed and amenable
to rational analysis while the input loading and the shear strength
properties of foundation and building materials under seismic
loading are at a primitive stage and little understood at this
time. It is felt that much computer effort has been diverted to
soil structure interaction based on guessed soil parameters and

that various in situ dynamic methods, such as shear wave
velocities, and dynamic response tests as well as well-designed
laboratory soil tests, are now being called for.

Another cause for concern is that there is now evidence to
show that during strong earthquakes ground accelerations have
been much larger than we used to think a decade ago. But there is
also evidence that many well designed and build structures, not
designed to resist earthquakes, survived these motions with
acceptable damage. This raises the awkward question of whether
current design techniques based on acceleration are reliable.
Here research may prove useful in attempting to develop design
techniques based on other physical parameters, such as velocity,
or energy flux.

The need for increasingly sophisticated dynamic analyses
often presents serious problems particularly in projects for
which survival conditions need to be assessed.

2.2.0. Vulnerability

Vulnerability may be defined as the degree of damage inflicted
by an earthquake of 'magnitude' (x) to man-made structures and
to the ground itself. It is a measure of the proportion of value,
as defined in equation (1), which may be expected to be lost as
a result of a given earthquake. Vulnerability, $v(x)$ can be
calculated in a probabalistic or deterministic manner for individual structures, if their dynamic behaviour to various earthquake
ground motions of 'magnitude' (x) is known, and can, by summation,
be estimated for a group of structures. The vulnerability of man-
made structures is the subject matter of earthquake engineering
and in contrast with Hazard it can be controlled and diminished
by appropriate engineering techniques, though sometimes at a cost
which must be justified by a reduced probability of loss.

For the assessment of vulnerability we rely mainly on field
observations. Every damaging earthquake affords an opportunity to
observe and study the damage caused by strong ground motions, and
there is already an abundant literature on the subject. These
studies, while making it possible to test the validity of certain
earthquake engineering hypotheses and techniques, have so far not
yielded much information permitting one to forecast the extent of
damage likely to be caused in the event of future earthquakes. The
principal reason for this is that only in very few cases has the
near-field or strong-ground motion (x) during earthquakes been
recorded by appropriate instruments. It has been generally impossible to correlate damage directly with strong ground motion and
thus to derive relationships of predictive value. Instead, we
have to resort to indirect correlations between $V(x)$ and (x)
using empirical attenuation laws of the type $(x) = f(R,M)$. The

results from such a correlation will depend on the suitability of
the attenuation law used and on the proper definition of the
vulnerability for different types of structures.

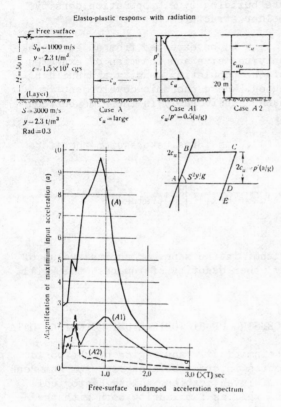

Figure 3. Elastoplastic
response of foundation
materials. Case A: elastic
response; Case A1: elastoplastic response of a
normally consolidated
deposit; Case A2: elasto-
plastic response of over-
consolidated deposit con-
taining a low-strength
layer.
Note foundamental diffe-
rences of surface response
spectra.
(see: Proc.3rd Europ.Symp.
Earthq.Eng., 1: 309,
Sofia, 1970)

Free-surface undamped acceleration spectrum

Quite often, damage is correlated with ground motions derived
indirectly from empirical relations with Intensity. Such a derivation is not suitable for risk assessment for two reasons. First,
the correlation between intensity (I) and maximum acceleration (x)
is very weak. It is not only that acceleration alone is not a
measure of damage, but also that Intensity is the mode, not the
median or mean of observed intensities within a particular area
that represents the value of the largest number of observations.
Consequently, Intensity should be treated as a probability distribution with a mean and a measure of its dispersion rather than
as a number. Second, Intensity (I) is, by difinition, a measure
of damage and consequently of V(x), where (x) is an independent
variable. Therefore (x) will cease to be an independent variable
once it is replaced by (I) and equation (1) will not any longer
be valid.

The use of Magnitude M as a measure of (x) seems to be promising. Appendix shows the results of a first order approximation
in the evaluation of the vulnerability of northeastern Mediterranean in terms of magnitude. This shows that the major controlling
factors of Vulnerability are building type, population density
and rate of ageing of the older structures.

Thus, a prediction of risk to be expected from an earthquake
must take account of not only the size and location of expected
events (Hazard) but also of the population density and building
types in the epicentral region. Where economic development is
rapid these factors can change considerably in the time between
comparable earthquakes.

Returning to equation (1), its formal expression therefore
should be:

$$\text{Risk} = R = \left[\int_{h_1}^{h_2} h'(x)V(x)dx + H(x_1) \right] \text{ X (Value)} \tag{2}$$

where Value may be taken either in the sense of capital value of
the production value or any other quantity of interest that will
give a dimension to (R).

3.0.0. COMPILATION AND PROCESSING OF HISTORICAL SEISMOLOGICAL DATA

Studies of long term seismicity in any region of the world
have to incorporate and process information acquired from numerous
and widely differing types of investigation. A long historical
perspective is desirable, as well as familiarity both with the
region itself and with the effects of earthquakes that have
occurred there. In addition to macroseismic information, instrumental data covering 20th century events should ideally be of
adequate high quality.

Only when these data are assembled can one begin to perceive
their long term significance. Analysis may be performed to elucidate
the vulnerability of certain areas, and evaluation of the distribution and time variation of the level of seismic activity over
a long period of time can also lead to a greater understanding of
the tectonic processes in the region and of the underlying laws
that may control the rate of activity. For the Eastern Mediterranean region long term observations may cover a period of about
20 centuries, for the Middle East 15 centuries, for North Africa
about 7 to 10 centuries and for Europe about 10 centuries.

3.1.0. Retrieval and analysis of macroseismic data

A region with a well-documented history may be capable of furnishing a considerable amount of information about earthquakes that have occurred in the past 10 to 20 centuries, defined here as the period before the advent of modern seismology. General chronicles, dynastic histories and archaeological findings may contain evidence of early events, and local histories in particular may describe the effects of earthquakes on monuments, local housing and the development of the area. These indications are generally restricted to well populated areas, particularly those in which a large urban centre is situated, and especially if the town has a long history of political and economic importance, as a trading centre, local administrative capital or political metropolis of the whole country. Information for such areas, and also for those less densely settled, may be provided by travellers passing through the country, or settling for a while in a town or area where they were interested to record local events.

Information about one country may be preserved not only by foreign visitors, but by diplomats on duty there, and other representatives, peaceful or belligerent, of foreign powers. Data found in foreign libraries and official archives may thus supplement indigeous information preserved in histories, private correspondence, or early and recent newspapers.

All these sources of information have to be identified and studied, to extend and correct any data listed in current earthquake catalogues, most of which are based on second or third hand information. The data they supply must be subjected to rigorous critical examination, not only to establish accurate dating and location of events in order to avoid duplication, but also in terms of the completeness of the sample. To this end, the characteristics of various historical records should be assessed, to see when and where the documents were being written, which areas were more important and which of more remote concern, which were populous and thus likely to suffer particularly from a large event, also to see which areas were frequently visited by travellers and which were not, along with the routes they took. How these factors influence the distribution of reported seismicity of the region must also be examined, so that allowances can be made for areas about which little information is available. These negative and positive factors can then be taken into account when an assessment is made of the level of seismic activity throughout the country over a long period of time. Under no circumstances should information be used at second-hand.

3.2.0. Calibration of macroseismic data

Once collected and assessed in the general terms outlined

above, it is then necessary to calibrate the data with observations
of recent earthquakes. This implies that adequate investigations
of recent events have been, or will have to be made, and that
field studies have concentrated on all relevant aspects of the
earthquake, such as damage to housing of different types, epicentral intensities, radii of perceptibility and ground effects.
Field studies may also bring to light information on earlier
events in the same region, particularly important if it is a
remote rural area that has not been prominent in documentary
sources.

As macroseismic data for recent events is likely to be more
full than it is for earlier earthquakes, to achieve a uniform
scale of assessment applicable to the whole period, it is necessary
to simplify the existing Intensity Scales and the criteria used
to quantify earthquake effects, concentrating on the salient
features that emerge from both historical and modern descriptions.
On this basis, historical data can be classified consistently
alongside modern data.

This stage being reached, one should be able to have assigned
epicentral intensities and radii of perceptibility to historical
and modern events on a uniform basis, making use of experience
from the field study of recent earthquakes. For obvious reasons,
this kind of work cannot be done by an arm-chair seismologist
or engineer. The effort involved will have included thorough
research in documentary sources, probably written in a wide
variety of languages, some retrieved relatively easily and some
with difficulty by reason of the inaccessibility, or the poor
state of preservation or legibility of the relevant material.
Secondary areas of study, such as archaeological work on sites,
and on the architectural history of both major monuments and
domestic dwellings, will have contributed some useful insights.
Geographical and historical study of the country will have been
carried out, particularly with regard to the distribution of
population and the channels of communication, which have influenced
the survival of data. In addition, thorough field studies will
have been made of as many earthquakes as possible. It will be
appreciated that these procedures take time and require a variety
of skills.

3.3.0. Retrieval and unification of instrumental data

To achieve a similar homogeneity for instrumental data, it
will be necessary to undertake a process of revising existing
epicentral location estimates and magnitudes, particularly for
events prior to 1960. This may be done from existing seismograms
and station bulletins, or it may be necessary, for events which
are important or not previously identified or processed, to resort
to original records. As with macroseismic information, the value

of instrumental data must be assessed and unified, in terms of
accuracy. The relocation of early instrumental epicentres is unlikely to result in new positions which are much reliable than the
originals, but major improvements can be achieved when these locations had only been adopted. For events after about 1960 and for
areas of limited geographical extent, macroseismic epicentres may
be used as calibration locations in joint epicentre determinations
(JED).

It will also be necessary to calculate magnitudes for all
events of the instrumental period using a standard procedure, viz.
M and m, as defined in the Manual of Seismological Observatory
Practice, making use of amplitude-period data from station bulletins or from seismograms. In cases where instrumental data are
lacking or inadequate for the calculation of magnitude, it may be
possible, for shallow depth earthquakes, to estimate magnitude
from other parameters, particularly the epicentral intensity of
the earthquake and its radius of perceptibility. As these can
generally be determined from macroseismic data, it is possible
to assign magnitudes to historical events.

The value of such a body of data, with a high level of internal consistency, is that further deductions can be made from it,
to explore the long term rate of activities in contiguous tectonic
elements and the time scales of seismicity transfer from one
element to another. Furthermore, it should be possible to discriminate regions of different types of seismic activity, in terms
of clustering or more uniformly distributed activity. This may be
of considerable value to the understanding of regional tectonics
and the identification of areas characterised by high seismic
activity in the past but which are currently quiescent.

Such a programme of research should therefore be of great
importance not only from a scientific but also from a practical
point of view. It will be of concern and interest to those responsible for planning large engineering structures and more modest
housing developments, and for attempting to minimise regional
vulnerability to earthquake hazard.

An important assumption in seismic hazard studies is the
choice of the upper bound limiting magnitude. A method of determining this magnitude is usually based on Gumbel's third type
asymptotic distribution which, however, requires that earthquakes
are produced by a stationary process, which does not seem to be
the case for most parts of the world studies. Long term observations suggest that in fact earthquake generating processes are
quasi-periodic or broadly clustered and also that much of the
tectonic deformation, quite often, occurs aseismically. An assumption of stationarity, therefore, would lead either to an over-
estimation or under-estimation of future activity, and this will

depend on whether the period of observation is taken from a period
of high or low local rate of seismicity. It is precisely this
assumption of stationarity, together with the use of short-term
data that had led seismologists to believe that earthquake
magnitudes in various parts of the world were unlikely to exceed
a certain calculated value, a limit recently surpassed by quite
a few shocks, viz. Barce in Libya of 21 February 1963 (M = 5.6),
Friuli in Italy of 6 May 1976 (M = 6.3), Montenegro in Jugoslavia
of 15 April 1979 (M = 7.2), El Asnam in Algeria of 10 October 1980
(M = 7.3), Sirch in Iran of 18 July 1981 (M = 7.0) and Aswan in
Egypt of 14 November 1981 (M = 5.4), to mention a few.

There is no doubt that the proper calibration of historical
events is essential for the assessment of risk.

Let us consider as an example the earthquake of 24 January
1927, one of the strongest shocks of this century to occur in the
North Sea area. It was felt over almost the whole of the south of
Norway, over much of Scotland, a small part of the North of England
and Northwest Denmark. Recent and old studies of this event, based
on data at second-hand, imply that in the U.K. this earthquake
was felt with an intensity VIII in the Modified Mercalli Scale,
and that it had a surface wave magnitude of 5.7. The following
summary, extracted from our files of historical seismicity, may
prove the value of the study of historical events.

3.4.0. The North Sea earthquake of 24 January 1927.

Various attempts to relocate the epicentre of this event
(Table I) suggest that its origin must be sought off the Norwegian
coast, about 150 kilometres west of Haugesund.

Accounts of the effects of this relatively small magnitude
earthquake in the U.K. and Norway were published by Tyrrell (1932)
and Kolderup (1930) respectively. Tyrrell's intensity distribution
map, Figure 4, shows that the maximum intensity in the U.K. reached
values of VII in the Peterhead district, Northeast Aberdeenshire,
about 350 kilometres from the epicentre, with a wide zone of intensity VI lying to the west of this district, that includes nearly
the whole of the Shetlands, Orkneys and Northwest, Central and
Southeast Scotland. For the eastern part of the affected area,
Kolderup's intensity distribution map, Figure 5, shows that intensities in Norway, in regions as close to the epicentre as 150 kilometres, did not exceed V, and on the average IV. The difficulty
in reconciling these two isoseismal maps is that the former, in
assessing intensities, has used the Rossi-Forel (RF) scale or
Davison's (DV) scale, which he considered to be equivalent, while
the latter had used Mercalli-Cancani's (MC) scale and a much larger body of observational records. The situation is complicated
with respect to the actual intensity scale used by Tyrrell for

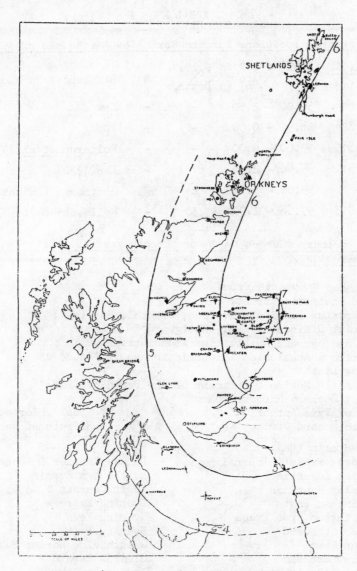

Figure 4. Isoseismals of the North Sea
earthquake of 24 January 1927;
(RF)-(DV) Scale, after Tyrrell (1932)

T A B L E   I

Epicentral locations for the North Sea Earthquake of 1927

	Origin Time	Epicentre N°	Epicentre E°	Focal Depth	M	References
1	051824	59.00	- 2.50	-	-	Kew
2	051824	58.50	- 6.00	-	-	ISS
3	051819	59.00	- 5.00	-	-	Kolderup et al(1930)
4	051811	59.40	- 2.90	(18)	-	Lee(1932)
5	-	59.00	- 3.00	n	$5\frac{3}{4}$	Gutenberg & Richter(1965)
6	051818	59.68	- 2.70	30	-	Neilson(1980)

List of primary sources used for the study of the North Sea earthquake 1927.

Unpublished Documents from:
  Clyde Navigation Trust*;
  Conservancy Commissions: Tees, Bristol, Harwich*;
  District Offices of H.M. Coastguard**
  Dock and Harbour Authorities Association*
  Fichiers & Annales de la Commission pour l'Etude des Raz de Maree*
  Lloyds List**
  Mercantile Marine Department of the Board of Trade*
  Mersey Docks & Harbour Board*
  Port of Tyne Authority*   * Tidal information
  Public Record Office, Kew**   ** Effects at sea.

Press Reports (UK):
  Aberdeen Press & Journal   Dundee Courier & Advertiser
  Berwick Advertiser   Edinbourgh Evening Dispatch
  Beverley Guardian   Elgin Courant & Courier
  Birmingham Post   Evening Express
  Bridlington Free Press   Evening News
  Border Standard   Evening Post
  Buchan News   Galashields Border Magazine
  Burnley Express   Glasgow Herald
  Cambrian Daily News   Inverness Courier
  Carlisle Journal   Newcastle Journal
  Cumberland Evening News   Orcadian
  Cumberland & Westmorland Herald   Orkney Herald
  Daily Express   Perthshire Advertiser
  Daily Mail   Scarborough Mercury
  Daily Record   Scotchman
  Daily Telegraph   Shetland Times
  Dumfries & Gulloway Courier   Strathearn Herald
                                    Times

Published Reports:

Dollar, A.(1951) "A catalogue of Scottish earthquakes 1916 - 1949"
        Trans.Geol.Soc.Glasg. 21:283-361,

Kolderup, C.(1930) "Jordskjelv i Norge 1926-1929"
        Bergen Museums Aarbok, 6:1-40,

Kolderup N., Krumbach G.(1930) "Det norsk-skotske jordskjelv 24-1-1927'
        Bergen Museums Aarbok, 7:1-16,

Lee, A. (1932) "The North Sea earthquake of 1927 January 24"
        Mon.Not.R.astr.Soc., Geoph.Suppl. 3:21-29

Neilson, G.(1980) "Historical seismicity of the North Sea"
        pp.119-37

Tillotson, E. (1969) "Notes on some recent earthquakes in the UK"
        Festschr.W.Hiller, Inst.f.Geoph.Stuttgart,108-25,

Tillotson, E. (1974) "Earthquakes, explosions and deep underground
        structure of the UK", Journ.Earth.Sci.pp.353-64, Leeds,

Tyrrell, G.(1932) "Recent Scottish earthquakes"
        Trans.Geol.Soc.Glasg. 19:1-41

Nature(1927) 119:208.

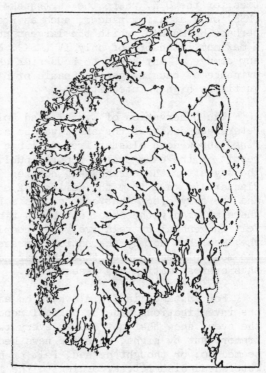

Figure 5. Intensity distribution in Norway of the
North Sea earthquake of
1927; (MC) scale, after
Kolderup (1931).

the U.K. by the fact that he equates his isoseismal V(FR-DV)
with II-0(MC) of Kolderup's and also that Tillotson (1969, 1974)
equates VII(RF-DV) with VIII on the Modified Mercalli (MM) scale.
A further complication, typical of the period, arises from the
fact that, in plotting their intensity distribution maps, both
Tyrrell and Kolderup have used maximum intensities rather than
intensities which represent the largest number of observations
at a given place. Tyrrell's data is rather limited and consists
of a number of accounts that he collected himself and of press
reports from two newspapers (The Glasgow Herald and Orcadian);
it does not seem that Tyrrell visited the region. Kolderup's data
consists of replies to about 360 questionnaires and of individual
reports.

To resolve the complication of reconciling the two different
isoseismal maps it was found necessary to evaluate anew intensities
throughout the affected area using a common scale. This was done
on the (MSK) scale which was employed to assess intensities from
a body of data much larger than that used by Tyrrell. For Norway,
the original data available consisted almost exclusively of newspaper reports. (see list of sources used)

Figure 6 shows the distribution of the re-assessed intensities for the U.K. where the earthquake reached an intensity of V
(MSK) only at a few places, such as Fraserburgh, Petershead and
perhaps Helmsdale, while the largest number of observations show
a maximum intensity of only IV (MSK). Nowhere did the shock cause
any damage of any kind, to houses or chimneys and there is no
evidence of causing general panic or other effects that could
justify a higher intensity.

The observation of Tyrrell and Dollar (1951) that in the
Petershead district the shock was so strong as to cause bells to
ring, throw down plaster from ceilings and clocks to stop is the
result of the exaggerated reports published in the Glasgow Herald
the day after the earthquake, not mentioned by the local press.
Plaster did fall from the ceiling of only one house at Petershead,
and the only clock that stopped was a French clock. At Aberdeen,
where allegedly the shock set bells ringing, the earthquake was
felt in general by relatively few people, a fact brought out by
the local but not by the national press used by Tyrrell, who for
the most places assumed that the earthquake was felt by everybody,
thus overestimating his assessments.

For instance in Elgin, where he assigns an intensity V-VI,
the investigations made by a local reporter in the morning after
the event show that not one in every twenty had been aware of the
tremor, but by night, when the news became general, everyone
seemed to, or thought he had, felt or heard something. Even in the
Petershead district, according to the local press, many people

Figure 6. Distribution of re-assessed intensities of the North Sea earthquake of 1927
in the U.K.

did not wake up, and most of those who did, did not realise it
was an earthquake. In Fraserburgh several men on night duty state
that they experienced nothing unusual. There is no doubt that
Tyrrell's intensities are grossly overestimated, even on the
(RF) scale, and that there are no grounds for considering
Tillotson's maximum intensity of VIII (MM) other than a misprint.

The shock was felt as far south as Scarborough. Lee (1932)
says that it was perceived also in Norfolk, but we could find no
evidence for this. At sea, the earthquake was felt on board a
vessel in Kirkwall Harbour (Orkneys) and offshore Helmsdale. With
the exception of a notice of great rollers observed 10 minutes

after the event at the bar of the mouth of the Helmsdale river,
there is no evidence of seismic sea waves. The tide curves for
the period 23rd - 25th January from the tide gauges at North
Shields (Tyne), Southend, Tilbury and London Bridge appear to be
quite normal. (Tidal information and graphs supplied by the Port
of Tyne Authorities (AWES/BH of 2.2.1977)).

Figure 7 shows the distribution of intensity throughout the
affected area. In Norway maximum intensities exceeded V (MSK) but
only at a few places, with the representative maximum intensity
that corresponds to the largest number of observations not
exceeding IV+ (MSK). Also shown in Figure 7 are the instrumental
epicentres calculated by various authorities, Table 1. Of these
determinations, Lee's and Neilson's (nos. 4 and 6) are probably
the best, lying less than 100 kilometres from the Beryl oilfield.

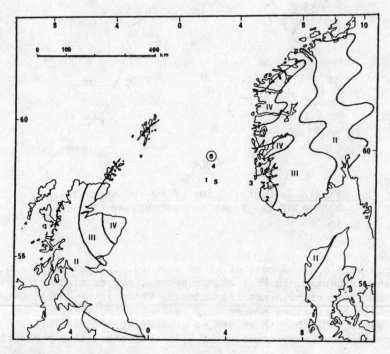

Figure 7. Distribution of intensities of the 1927
North Sea earthquake.

The surface wave magnitude of the earthquake, calculated from readings of 20 stations, using the Prague formula without station corrections, was found to be 5.3 ± 0.3; with station corrections this value becomes 5.2 ± 0.4. The normalised magnitude at 20 seconds period, as defined by Marshall and Basham, is 5.1 ± 0.3. The seismic moment of the event was found difficult to calculate. Preliminary results indicate $M_o = 4 \times 10^{23}$ dyne-cm.

4.0.0. DEFICIENCIES IN EXISTING FACILITIES IN NORTH WESTERN EUROPE

Deficiencies in existing facilities in North Western Europe have been identified as:
(1) Insufficient expertise in tectonics and Quarternary Geology.
(2) Lack of sufficient exchange of knowledge between seismologists, geophysicists, geologists and the civil engineering industry.
(3) Insufficient coordination within the seismological services (National Institutes, Universities, Research Centres, Industry).
(4) Lack of readily available, reliable, up to date information on hazard assessment.
(5) Insufficient training of engineers and other design professionals in seismic safety.

The assessment of seismic hazard and the behaviour of foundation materials under seismic loading are two fields that are likely to become of great importance. The siting and designing of nuclear plants, petrochemical installations, fuel disposal units, offshore structures and submarine pipelines will be controlled by the degree of their vulnerability and the seismic hazard of the region. The monitoring of engineering structures to vibrations and displacements becomes then an important consideration.

Further research and development should be focussed on improved understanding of the behaviour of soils and rock under dynamic loads. In order to achieve this, better laboratory and field instrumentation, including shaking-table facilities, has to be developed and tested. Northwestern European countries have historically an excellent record in the field of static soil mechanics and very little extra effort would be required to extend this into the dynamics field. A clear lead in this field would not only help solve problems in these countries but would attract overseas projects.

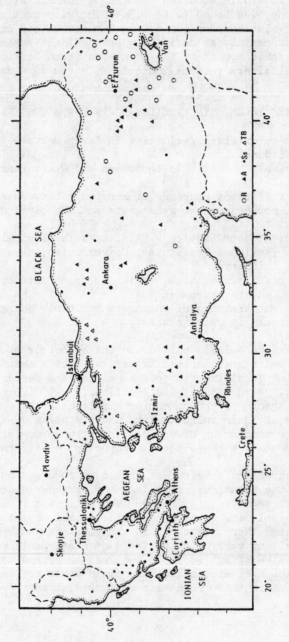

Figure 8. Location map of study region showing distribution of predominant building types associated with earthquakes during the period 1902 – 1981. (see:Disasters, 5:4, London 1981)

APPENDIX

The purpose of this appendix is to show how previous interdisciplinary experience within a region can be used to assess
future earthquake damage.

The region chosen for study lies between 34 and 43 degrees
north and 19 and 44 degrees east, Figure 8. Within this highly
seismic region, which includes Greece and Turkey, large shallow
(c. 15 km) and intermediate (> 70 km) depth earthquakes affect a
variety of building types distributed over an area of widely
varying population density. The area also permits the assessment
of damage from offshore earthquakes, which is an important
consideration for areas bordering the North Sea.

In order to assess potential seismic damage in a region,
information is needed on which the physical earthquake parameters,
the population density and the building types can be based.

Figure 9 shows a plot of damage D versus epicentral intensity
I_o and the estimated regression line for shallow earthquakes with
epicentral regions entirely on land. The qualitative conclusion

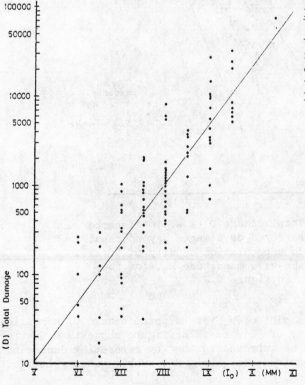

Figure 9. Total damage
(D) (number of houses)
versus epicentral
intensity (MM) for
events occurring in
the region shown in
Figure 8.

that can be drawn by observing this plot is that I_o is not a good predictor of D. This is obvious, since two shallow depth earth-quakes of different magnitude may well produce the same epicentral intensity, however, the larger of them will release more energy that will affect a larger epicentral area and consequently cause greater total damage D.

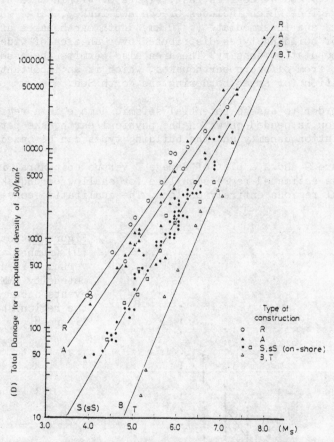

Figure 10. Total damage D (i.e. total number of houses destroyed or damaged beyond repair) normalised to a population density of 50/km², versus surface-wave magnitude M_s, for the region shown in Figure 8.

Figure 10 shows a plot of M_s versus total damage D, normalised to a population density of 50 / km². The figure suggests a distinct correlation which can be improved further by separating damage to different types of construction, i.e.: R = rubble masonry houses;

A = adobe-brick constructions; s or S = stone masonry dwellings
and T or B = timber-braced or timber-framed brick houses with
relatively light roofs, Figure 10. The conclusion that can be
deduced from this Figure is that the type of construction over-
shadows all other variables. Thus for a population density of
$50/km^2$, an earthquake of a small magnitude, say 4.0 will render
uninhabitable 1 and 200 houses of the two extreme types of
construction (R, A and T, B) respectively, if these structures
happen to be within the epicentral area of the shock.

Figure 11 shows the percent total loss as a function of
magnitude for the four types of construction. These curves depict
the high vulnerability of the substandard classes of construction
R and A and the higher resistance of classes s and S. This figure
also explains why intensities derived from values of percent of
total loss or damage can be grossly overestimated or underestimated
if the appropriate class of man-made structures exposed to the
earthquake is not taken into consideration.

Figure 11. Percent of
total loss of houses
exposed to an earth-
quake of surface-wave
magnitude M_s.

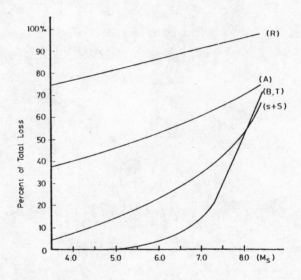

SEISMIC RISK AND THE NORTH SEA[*]

Paul W. Burton[1], Ian G. Main[1,2] and Graham Neilson[1]

[1]Institute of Geological Sciences, Murchison House,
West Mains Road, Edinburgh EH9 3LA
[2]University of Edinburgh, Department of Geophysics,
Mayfield Road, Edinburgh EH9 3JZ
[*]Presented by Dr. P.W. Burton

ABSTRACT

There is considerable scope for improving knowledge
pertinent to the assessment of seismic risk in the North Sea,
and evaluation of uncertainties on estimates of the contemporary
seismic risk is vital.

Gumbel III extreme value statistics indicate a 200-year
earthquake of magnitude 6.3 ± .5, and an uncertain largest
earthquake magnitude ω = 6.7 ± 1.9. Combining these recurrence
statistics with attenuation laws shows that the earthquake most
likely to be felt at a point at any intensity has magnitude
about 6.2, whereas the most perceptible earthquake in terms of
ground acceleration is about 5.2: neither intensity nor acceleration alone are sufficient to represent the complete earthquake
dynamics of seismic risk.

Annual exceedance probabilities at a point corresponding to
peak ground acceleration of .05g and .20g are in the ranges
1.11-1.42 x 10^{-3} and 0.53-1.40 x 10^{-4} respectively, variations
depending on the variety of attenuation laws chosen and focal
parameter assumptions, and may be scaled by alternative
estimates of the North Sea area km^2.

INTRODUCTION

The major purpose of this paper is to discuss some aspects
of seismic risk which pertain in the North Sea, which for this
purpose will be taken as extending to and including the
Norwegian Sea. Firstly, different data bases available for the

347

A. R. Ritsema and A. Gürpınar (eds.), Seismicity and Seismic Risk in the Offshore North Sea Area, 347–364.
Copyright © 1983 by D. Reidel Publishing Company.

North Sea will be briefly discussed, and attention will be drawn
to the inadequacies of these data. In the light of these shortcomings it may be possible to suggest subject areas where
further studies may be most pertinent to the numerate assessment
of risk. Despite inadequacies in available data, numerate
estimates of seismic risk in the North Sea will be calculated,
with some attention to the identification of quantified uncertainties in the estimates, and this analysis will be the main
emphasis of the paper. Seismic risk will be understood herein
as the probability of occurrence of different levels of earthquake magnitude, and this will be extended to the probability of
different levels of earthquake produced ground motion at any
site within the North Sea.

INFORMATION PERTINENT TO NORTH SEA SEISMIC RISK

A seismic risk analysis involves consideration of several
component aspects [1], amongst which may be included:

1.   Seismo-tectonic model: a spatial distribution of
      earthquake epicentres within a tectonic framework.

2.   Earthquake catalogue: a chronological list of earth-
      quake occurrences with some parameters relating to the
      energy, size or damage potential of the earthquake;
      for example magnitude and intensity.

3.   Attenuation law: a means of quantifying the decrease
      of seismic energy with distance from the epicentre or
      capable fault.

4.   Statistical framework: a means of quantifying the
      probabilities of hazardous occurrences, along with the
      corresponding uncertainties, given the fundamental
      geophysical and seismological data for the area
      concerned.

5.   Earthquake dynamics: the dynamics of ground vibration
      caused by earthquakes (strong motion) and its inter-
      action with engineering structures.

It is immediately apparent that geotechnical data and local soil
amplification factors are not directly included in this scheme,
nor is earthquake strong motion data, as typified by accelerograms, beyond the extent to which these impinge on attenuation
laws. Equally, earthquake dynamics and engineering design is
outside the scope of this paper. The remaining aspects can be
considered briefly.

A seismo-tectonic model of the North Sea hardly exists, at

best it is rudimentary. However, it is important to distinguish
between a tectonic model and a seismo-tectonic model of the
North Sea. There is a considerable amount known about the
tectonics, to which the work and substantial reviews of Kent
[2], Ziegler [3] and others bear testimony. Never-the-less the
tectonics of the North Sea is far from being a closed issue, as
discussion on the extensional origin for grabens in the North
Sea illustrates [4,5]. Even though tectonic knowledge of the
North Sea is considerable, these debates on both detail and
fundamental issues seem bound to continue. Perhaps the most
perplexing difficulty with such detailed tectonic models is the
likelihood that a posteriori explanations of earthquake occurrence may readily be achieved, whereas a priori prognostications
are not forthcoming from the same models.

Seismo-tectonically the North Sea is in an intraplate
situation, and therefore remote from the major earthquakes more
typically associated with interplate earthquakes at plate
boundaries. The raison d'etre of intraplate earthquakes is
regarded as a difficult problem, as evidenced by Sykes' [6] substantial work. Intraplate earthquakes do not often exhibit a
systematic spatial pattern, which is largely true of much of the
seismicity associated with the North Sea. It seems, therefore,
that there is little to suggest that significant earthquakes may
not occur at any place beneath the North Sea. This is different
in its implications to the more obvious statement that
significant earthquakes may be felt at any place within the
North Sea.

Most of our knowledge of historical earthquakes con-
tributing to the seismic risk in the North Sea comes from macroseismic evidence, felt reports from the surrounding land masses.
Many countries and seismologists around the North Sea have
contributed to this knowledge, and various earthquake catalogues
have emanated from the North Sea countries. Examples of macroseismic earthquake catalogues are: for Europe, Karnik [7,8]; for
Britain, O'Reilly [9], Davison [10], Dollar [11] and Tillotson
[12]; for Fennoscandia, particularly Norway, Keilhau [13],
Kolderup [14] and Båth [15].

There has to date been little attempt to reconcile or unify
these earthquake catalogues for the offshore areas, where earthquake epicentres may only be inferred due to the nature of
macroseismic data. It is also possible that only the larger
earthquakes are likely to be noted. For reasonably well
controlled macroseismic epicentres inferred near to the median
line between the land masses implies a large radius of
perceptibility so that the earthquake is felt within countries
either side of the North Sea. This is borne out by our macroseismic knowledge of some significant North Sea earthquakes

during this century. These are the 1904 October 23 Oslofjord
earthquake with magnitude M = 6.4; 1927 January 24 off the west
coast of Norway with magnitude M = 5.1 (see for example Tyrell's
[16] macroseismic evidence); 1931 June 7 Dogger Bank with M =
6.0 (see Versey [17]). Even for these latter two significant
earthquakes uncertainties in epicentral location may be significant, and the advantages of even the early instrumental era in
re-locating these earthquakes are described in detail by Neilson
[18]. Papastamatiou [19] has also demonstrated how useful
physical source parameters may be obtained from the few records
of the 1904 event, and additionally, how instrumentation of that
era was perhaps in some ways appropriate to our present needs.
Although additional information can be obtained from these
instrumental data, there is the basic identifiable need to
reconcile the macroseismic data with existing instrumental data,
and in due course with the growing body of microseismic data
[20] nearer to the North Sea earthquakes. A magnitude scale
applied uniformly to the North Sea and the surrounding
territories, which related directly to the international
standards [21] without an unknown regional bias [22], would be
most useful. Although no such ideal catalogue exists, the preliminary catalogue cited by Ritsema [23] will suffice here.

Attenuation of seismic energy in the North Sea, judging
from descriptions of perceptible radii or areas for earthquakes
of reasonably accurately known magnitude, seems to be low.
Attenuation laws for intensity are not well documented, as would
be expected for an offshore area, and it is therefore reasonable
to seek an acceptable intensity attenuation law from a
comparable region or nearby country. This procedure will be
adopted here. Earthquake strong motion data - accelerograms -
are not available for the North Sea or this part of north west
Europe in general. Attenuation of the ground motion parameter
x with focal distances R km is typically expressed in the form:

$$x = a \ e^{bM} R^{-c} \tag{1}$$

where M is the earthquake magnitude and a,b,c are constants.
When indigenous strong motion data is not available, it is
necessary to use a 'comparable' law from elsewhere. Several
mooted laws rely on Californian data, a few on European Alpide
belt data. An average law representative of several of those
available has been proposed by Makropoulos [24], and this will
be invoked in the subsequent analysis. A case remains for
designing experiments in the North Sea which measure geo-
physically valid parameters of seismic absorption rather than
using ad hoc formalisms like equation (1). Measurement of in
situ attenuation or seismic absorption in the North Sea would
probably generate reduced uncertainty in the ensuing risk
calculated by statistical methods.

The statistical methods available allow calculation of the seismic risk, including the vital possibility of carefully assessed uncertainty in those estimates. Deterministic methods will not be included here. Although the family of statistical distributions is large, three sub-families will be mentioned here. If it is assumed that earthquakes are causally connected events then the use of Markov chains is appropriate, and a relatively smaller group of researchers advocates and is developing this approach. Causally unconnected earthquakes is the more usual assumption, which further divides into two groups: whole process and part process statistics. The whole process implies that details of all earthquakes down to a particular magnitude threshold must be known in order to determine a frequency-magnitude law. The part process requires only that extreme values of the earthquake occurrence process is known. The latter has its attractions, particularly for the North Sea, where magnitude detection thresholds have not been uniformly low and it is likely that our knowledge of the extreme events is more complete. Gumbel's [25] extreme value theory will be used here, including the additional advantages of Gumbel's third asymptotic distribution of extreme values, considered promising by Bath [21], which incorporates an upper bound to earthquake magnitude occurrence in the distribution, that is, the maximum credible magnitude or largest earthquake is taken into account by the statistical process. This is an essential feature, although it will become apparent that it is the forecasting capability of the distribution which is important rather than the precision to which the upper bound magnitude is known at vanishingly small probabilities of occurrence.

SEISMIC RISK ANALYSIS

Earthquake Magnitude Occurrence.

The Gumbel III distribution is of the form:

$$P_n(M) = \exp\left[-\left(\frac{\omega - M}{\omega - u}\right)^{\frac{1}{\lambda}}\right] \qquad (2)$$

having three parameters $[\omega, u, \lambda]$ which are respectively ω the upper bound to magnitude M, u the characteristic extreme magnitude value associated with the n-year extreme interval, and λ a measure of the distribution curvature. The probability that magnitude M is an extreme value is $P_n(M)$, and $P_n(M)$ may be related to annual extremes using $P_1{}^n = P_n$. Full details of the application of equation (2) to seismic risk in Southern Europe are given in Burton [26], including a full description of the covariance matrix ϵ among parameters $[\omega, u, \lambda]$.

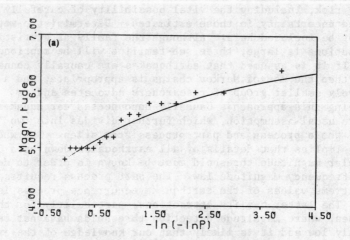

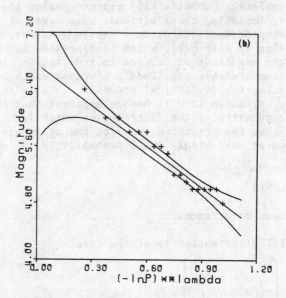

Fig. 1. Three-year interval extreme magnitudes (crosses) of North Sea earthquakes for 1900–1979 with the Gumbel III extreme value distribution fitted to these data. P is the probability that a magnitude M occurrence is an extreme over a three-year interval.
(a) The standard plot with M against −ln(−lnP).
(b) Plot of M against $(-\ln P)^{\lambda}$ which results in a linearised plot, the upper bound magnitude ω being the intercept on the M axis. The envelopes represent ± σ.

Using Ritsema's [23] preliminary catalogue for 1900 onwards, three-year extreme intervals, a magnitude threshold equivalent to M = 4.75, an uncertainty in each extreme magnitude assumed to be 0.5, and fitting equation (2) to these data, generates an $[\omega,u,\lambda]$ parameter suite of:

$$\omega = 6.70 \pm 1.9$$
$$u = 4.82 \pm 0.26$$
$$\lambda = 0.35 \pm 0.56$$

with an associated covariance matrix ε:

$$\begin{bmatrix} 3.484 & 0.240 & -1.016 \\ 0.240 & 0.067 & -0.089 \\ -1.016 & -0.089 & 0.311 \end{bmatrix}$$

These data and fitted curve are plotted in Fig 1(a) which shows curvature towards the upper bound magnitude ω. The linearised plot in Fig 1(b), which is the preferred presentation, would intercept the ordinate axis at the value ω. Note the large uncertainty in ω (obtained from σ_ω^2 in ε). There is also a large negative covariance between ω and λ (obtained as $\sigma_{\omega\lambda}^2 = -1.016$ in ε), which becomes of significance when uncertainties in the forecasting of different magnitude levels are calculated using $[\omega,u,\lambda]$.

$[\omega,u,\lambda]$ can be used to forecast M(T) for different T-year earthquake magnitudes using equation (2) rearranged to:

$$M = \omega - (\omega-u)\left[-\ln(P_1(m))\right]^\lambda \tag{3}$$

Forecasts of M(T) for the North Sea contrasted with m(T) for the British land mass excluding offshore earthquakes [27] are:

T-years	North Sea M(T), σ_M	Britain m(T), σ_m
10	5.40, 0.16	4.62, 0.12
25	5.81, 0.18	5.05, 0.12
50	6.02, 0.23	5.20, 0.16
100	6.17, 0.34	5.29, 0.21
200	6.30, 0.50	5.35, 0.26
∞	6.7, 1.86	5.5, 0.4

It must be emphasized that M(T) approximates to the surface wave magnitude M_s and m(T) the body wave magnitude m_b, which scales typically cross over at about magnitude 5.25. These results based on the limited historical earthquake catalogue available suggest that the North Sea is capable of significantly greater earthquakes than the British land mass.

Largest earthquake (maximum credible earthquake) The upper
bound magnitude or largest earthquake commensurate with these
data is not precise, being in the vicinity of M = 6.7. Despite
the uncertainty on this figure it cannot be ignored in an intraplate situation where the seismicity does not exhibit a clear
spatial pattern and 'rogue' earthquakes |26| may occur. This
value is generated from an earthquake catalogue of short time
span relative to the long return periods or low annual probabilities it is wished to forecast. These results can be contrasted with the intraplate central and eastern USA,which in
many other respects is similar to the North Sea, but is note-
worthy because it has experienced earthquakes of magnitude in
excess of 7 during the last few hundred years. Particular
examples are the New Madrid earthquakes of 1811/12 and that in
Charleston, South Carolina, in 1886. Main and Burton's |28|
analysis of earthquake catalogues encompassing the era of these
earthquakes gives an upper bound magnitude in terms of a
macroseismic body wave magnitude scale, which would not be
expected to saturate, of ω = 7.7 ± 1.2. This is considerably
higher than the North Sea.

Perceptibility and Partitioned Risk

      The previous section has concentrated on the prediction or
forecasting of the frequency of different levels of earthquake
magnitude, that is magnitude recurrence where magnitude is taken
as an absolute measure of earthquake size, energy or destructive
potential. It is also necessary to predict the probability of
different levels of ground vibratory motion x being exceeded,
where x may be taken as intensity, peak ground acceleration or
velocity, RMS ground acceleration etc, as the earthquake
engineer requires.

      Following Burton |27,29| perceptibility P_p is defined as
the probability that a point perceives ground vibratory motion
of level x or higher when an earthquake magnitude M occurs,
that is:

$$P_p(x \mid M) = P_c(x) \cdot \phi(M) \tag{4}$$

where $\phi(M)$ is the probability density function of the earthquake
frequency-magnitude law, and $P_c(x)$ estimates the probability
that a point in the North Sea is within the area which perceives
ground motion at least at level x when M occurs. The latter is
approximated by the ratio of the felt area corresponding to x
divided by the area of the North Sea, taken to be Ritsema's |23|
value of 255.10^4 km^2. Should other 'areas' be invoked as
corresponding to the North Sea, then all the following values of
risk must be appropriately adjusted.

      Perceptibility $P_p(x \mid M)$ effectively generates values of
partitioned risk, that is, the ensuing diagrams reveal the

potential contribution of each magnitude M to the probability of experiencing each level of ground motion x or higher at any point. The calculation of perceptibility requires an attenuation law.

Intensity perceptibility $P_p(I \mid M)$ is shown in Fig. 2(a) where areas perceiving I or greater, required in equation (4), have been derived from Lilwall's [30] macroseismic attenuation law for the UK:

$$I = 3.1m_b - 2.5 - 1.91nd \tag{5}$$

$$d^2 = h^2 + r^2 + k^2$$

and h km is focal depth, r km is surface distance and k (=20) corrects for finite focal volume. Intensity perceptibility is shown for a suite of intensity values, and it is seen that the resulting curves are nested. The nominal zero depth earthquakes show peak perceptibility at one and the same magnitude for all levels of intensity; this is the 'most perceptible earthquake' $M_p(I)$. Finite values of focal depth depress perceptibility at all intensity levels, and move M_p slightly towards the higher magnitudes at the right of the distribution.

Acceleration perceptibility $P_p(a \mid M)$ has similar properties, and is shown in Fig. 2(b) derived using a Cornell [31] attenuation law:

$$a = 2000 \ e^{0.8M} \ (r^2 + h^2 + k^2)^{-1} \ cm/sec^2 \tag{6}$$

$P_p(a \mid M)$ may also be derived using Makropoulos' [24] attenuation law:

$$a = 2164 \ e^{0.7M} \ (R + 20)^{-1.8} \ cm/sec^2 \tag{7}$$

where R km is focal distance. This generates results which are in many ways similar to Fig. 2(b).

Comparing Figs. 2(a) and 2(b) shows that the perceptibility curves based on intensity are sharper than those based on peak horizonal ground acceleration. Secondly the most perceptible magnitudes $M_p(I)$ and $M_p(a)$ occur at significantly different magnitudes $M_p(I) = 6.2$, and $M_p(a) = 5.2$ for both equations (6) and (7). Analysis using peak ground velocity might generate other values. Single valued seismic design criteria are not be to expected and the onus is on the earthquake engineer to consider the implication of the options presented to him.

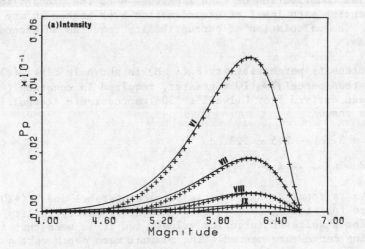

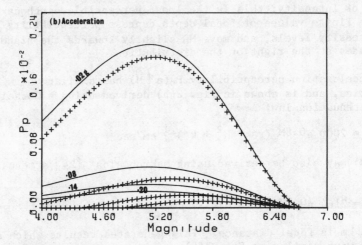

Fig. 2 Perceptibility combines magnitude recurrence statistics
with an attenuation law, and is the conditional annual
probability at a point of a given ground motion or larger
associated with an earthquake magnitude M occurrence.
(a) Intensity perceptibility $P_p(I)$ at a point is the annual
probability of Modified Mercalli Intensity I or higher
associated with an earthquake magnitude M occurrence.
(b) Similarly, $P_p(a)$ is acceleration perceptibility as a
fraction of g. In (a) and (b) solid lines represent notional
surface foci, with no focal size correction; crosses represent
foci in the crust, notionally at 10km depth, with a 20km
correction for focal size.

Integrated Perceptibility

By integrating equation (4) throughout the magnitude range
the annual probability of perceiving different levels of ground
motion at a point in the North Sea is obtained:

$$P_{ip}(x) = \int_{-\infty}^{\omega} P(x \mid M) \, dM \tag{8}$$

Resulting values of $P_{ip}(I)$ for intensity are shown in Fig. 3(b)
for both notional surface foci and shallow foci suites of earthquakes. Figure 3(a) shows the partitioned risk or perceptibility being integrated through M up to ω; values of $P_{ip}(I)$
being obtained at $M = \omega$. The values obtained for $P_{ip}(I)$, annual
probability of intensity I or greater at a point in the North
Sea, are:

$$P_{ip}(I).10^{-3}$$

Modified Mercalli Intensity	Surface foci	Shallow foci
VI	5.43	5.14
VII	1.90	1.68
VIII	0.67	0.51
IX	0.23	0.14

Similarly, annual probabilities of peak ground acceleration
a or greater at a point in the North Sea may also be calculated
using either acceleration attenuation law. Results for $P_{ip}(a)$
are:

$$P_{ip}(a).10^{-3}$$

Acceleration, g	1	2	3	4
.02	4.11	4.01	4.51	4.42
.05	1.42	1.33	1.11	1.01
.08	0.75	0.66	0.47	0.38
.11	0.45	0.38	0.24	0.17
.14	0.29	0.23	0.14	0.085
.17	0.20	0.15	0.082	0.044
.20	0.14	0.097	0.053	0.023
.23	0.097	0.065	0.038	0.012
.26	0.070	0.045	0.029	0.064

where columns 1 and 2 are derived using equation (6) for surface
and shallow foci earthquakes respectively, and columns 3 and 4
similarly correspond to use of equation (7).

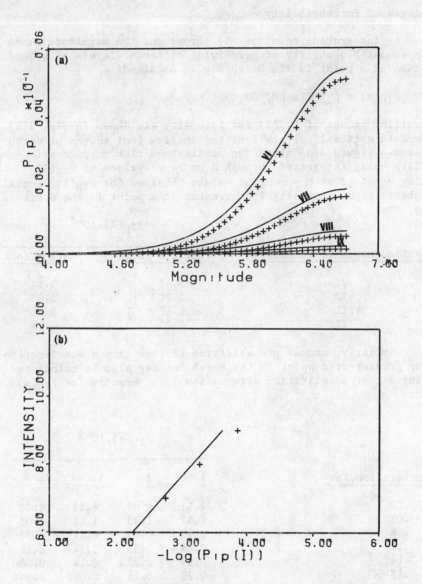

Fig. 3. (a) Integrating intensity perceptibility curves of
Fig 2(a) through magnitude up to M_1 generates annual
probability at a point of intensity I or greater, associated
with all earthquakes up to magnitude M_1.
(b) $P_{1p}(I)$ is the annual probability at a point of intensity I
or greater. It is obtained from the curves in (a) when the
integration reaches the upper bound magnitude at $M_1 = \omega$. Solid
curves and crosses as in Fig. 2.

Uncertainty in these risk calculations attributable to imprecise knowledge of seismic attenuation is shown in Fig. 4, where the above results obtained from two different laws are plotted. Alternative attenuation laws can also be debated as more relevant to the North Sea area, for example, the combined work of Ambraseys [32] and Chiarrutini and Siro [33] which derive results from European and Alpide strong motion data.

RECOMMENDATIONS AND CONCLUSIONS

The major concern is evaluation of the contemporary seismic risk in the North Sea, and its potential consequences for off-shore structures. Concern is necessarily with events having low levels of annual probability. It is necessary to understand if the damage potential of earthquakes to offshore structures changes rapidly or slowly with the seismic risk probabilities, rather than assess it at one arbitrary annual probability or return period. These calculations ideally require improved earthquake catalogues.

There are different aspects to seismic risk: there is the question of magnitude recurrence, and the ensuing problem of ground motion perception. The latter aspect can itself be divided into partitioned risk where the contribution of specific magnitude earthquakes to different levels of ground motion can be identified, and secondly into the overall calculation of possible levels of ground motion not associated with any partic-ular earthquake type. The earthquake engineer should be aware that a choice of seismic design criterion based on one ground motion parameter chosen from intensity, peak ground acceleration or velocity etc need not be representative of the entire physical process of earthquake dynamics. These calculations require attenuation laws on the seismologists part, and an understanding of the implications of different ground motion parameters to structures on the earthquake engineers part.

Inconclusive knowledge of North Sea seismic risk is con-tributed to by deficiencies in precise knowledge on several facets of the analysis, and the following emerge as desirable:

1.   Improved seismo-tectonic modelling, with the emphasis firmly on a priori prognostication of earthquake capability, or otherwise, of the North Sea tectonics.

2.   An improved unified earthquake catalogue, using existing data available throughout the countries around the North Sea.

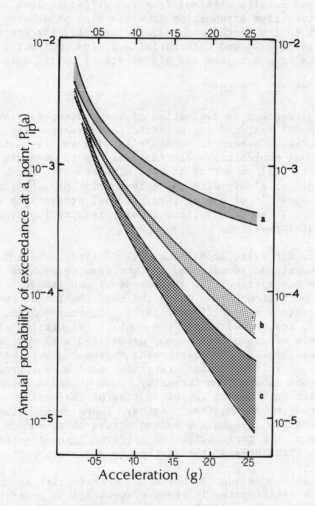

Fig. 4. Annual probabilities of earthquake accelerations at a
point in the North Sea. Variation arises from the choice of
attenuation law and earthquake parameter focal assumptions.
(a) Values spanned using Cornell's and Makropoulos' attenuation
with notional surface foci uncorrected for focal size.
(b) Values spanned using Cornell attenuation and notional
crustal foci, with 20 km focal size correction, and foci
notionally at 0 and 10 km depth (upper and lower extents of
speckled area respectively).
(c) As for (b) but using Makropoulos' attenuation.

3. Evaluation of seismic attenuation, possibly through
    controlled experiments designed to measure geophysical
    parameters of energy dissipation.

4. Improved evaluation of the uncertainties on seismic risk
    estimates, and the evaluation of seismic risk for a suite
    of risk levels rather than at an arbitrarily chosen
    probability level.

5. Awareness amongst seismologists and earthquake engineers
    that single ground motion parameters for seismic risk
    design criteria are not representative of the entire
    earthquake dynamic process.

Specific conclusions arising from the foregoing analysis of
a data base with acknowledged inherent limitations are:

1. Extreme value statistical analysis is effective for the
    North Sea.

2. The earthquake with a 50-year average repeat time has
    magnitude 6.0 ± .23, for 200-years the magnitude is
    6.3 ± .5.

3. The largest or maximum credible earthquake has magnitude
    $\omega = 6.7 \pm 1.9$. This value has large uncertainty and is
    obtained as a by-product inherent to the application of
    Gumbel III extreme value statistics. This earthquake is
    larger than onshore seismicity suggests for the British
    land mass, and is smaller than the earthquake suggested by
    a similar analysis for the eastern USA.

4. Not withstanding the vagaries introduced by using attenua-
    tion laws which have not been explicity calculated for the
    North Sea; the most perceptible earthquake magnitude using
    intensity as the measure of ground motion is $M_p(I) = 6.2$,
    and using peak ground acceleration it is $M_p(a) = 5.2$. One
    ground motion parameter alone is not sufficient to repre-
    sent the entire seismic risk and earthquake dynamic
    process.

5. The evidence from two peak ground acceleration attenuation
    laws suggests that the annual probability of exceedance at
    a point of .05g and .20g is in the ranges $1.11 - 1.42 \times 10^{-3}$ and $0.53 - 1.40 \times 10^{-4}$ respectively for the assumption
    of surface focus earthquake with finite focal size.
    Introducing knowledge of earthquake focal depths, when
    available, will reduce these values of seismic risk.
    Alternative values for the areal extent of the North Sea
    used in these calculations will scale these values.

ACKNOWLEDGEMENTS

    This work was supported by the Natural Environment Research
Council and is published with the approval of the Director of
the Institute of Geological Sciences (NERC).

REFERENCES

[1] Burton, P.W., 1978: Assessment of seismic hazard in the
    UK, in "Instrumentation for ground vibration and
    earthquakes",Institution of Civil Engineers, London, pp.
    35-48.

[2] Kent, P.E., 1975: Review of North Sea Basin development,
    J. geol.Soc.Lond., 131, pp. 435-468.

[3] Ziegler, P.A., 1975: Geologic evolution of North Sea and
    its tectonic framework, Bull.Am.Ass.Petr.Geol., 59, pp.
    1073-1097.

[4] Christie, P.A.F. and Sclater, J.G., 1980: An extensional
    origin for the Buchan and Witchground Graben in the North
    Sea, Nature, 283, pp. 729-732.

[5] Smythe, D.K., Skuce, A.G. and Donato, J., 1980: An
    extensional origin for the Buchan and Witchground Graben in
    the North Sea: geological objections, Nature, 287, pp.
    467-468.

[6] Sykes, L., 1978: Intraplate seismicity, reactivation of
    preexisting zones of weakness, alkaline magmatism, other
    tectonism postdating continental fragmentation, Revs.
    Geophys. & Space Phys., 16, pp. 621-688.

[7] Karnik, V., 1968: Seismicity of the European Area, Part I,
    Academia, Praha.

[8] Karnik, V., 1971: Seismicity of the European Area,
    Part II, Academia, Praha.

[9] O'Reilly, J.P., 1884: Catalogue of the earthquakes having
    occurred in Great Britain and Ireland during historical
    times: arranged relatively to localities and frequency of
    occurrence to serve as a basis for an earthquake map of the
    three kingdoms, Trans.R.Irish Acad., 28, pp. 285-316.

[10] Davison, C., 1924: A history of British earthquakes,
    Cambridge University Press.

[11] Dollar, A.T.J., 1949: Catalogue of Scottish earthquakes,

1916–1949, Trans.Geol.Soc., Glasgow, 21, 1, pp. 283–361.

[12] Tillotson, E., 1974: Earthquakes, explosions and the underground structure of the United Kingdom, J.Earth Sciences, Leeds, pp. 353–364.

[13] Keilhau, B.M., 1836: Efterretninger om Jordskjaelv i Norge Mag. for Naturvidenskaperne, 12, pp. 82–165.

[14] Kolderup, C.F., 1913: Norges jordskaelv, Bergens museum Aarbok 1913, 152pp. (In Norse).

[15] Bath, M., 1956: An earthquake catalogue for Fennoscandia for the years 1891–1950, Sveriges Geologiska Undersokning, Ser C., No.545, Arsbok 50 No 1, 52pp.

[16] Tyrrell, G.W., 1932: The North Sea earthquake Jan 24, 1927, Trans. Geol. Soc. Glasgow, XIX, pp. 10–29.

[17] Versey, H.C., 1939: The North Sea earthquake of 1931 June 7, M.N.R.A.S. Geophys. Suppl., IV, pp. 416–423.

[18] Neilson, G., 1979: Historical seismicity of the North Sea, in "Energy in the Balance" (Westbury House, publisher), Annual Meeting of the Brit.Ass.Adv.Sci., Edinburgh, pp. 119–137.

[19] Papastamatiou, D., 1978: Contributions of early instrumental seismic recordings to engineering analysis, in "Instrumentation for ground vibration and earthquakes", Institution of Civil Engineers, London, pp. 119–124.

[20] Browitt, C.W.A., and Newmark, R., 1981: Seismic activity in the North Sea to April 1981, Inst. Geol. Sci., Glob. Seism. Unit Report No. 150.

[21] Bath, M., 1981: Earthquake magnitude – recent research and current trends, Earth–Science Reviews, 17, pp. 315–398.

[22] Chung, D.H. and Bernreuter, D.L., 1981: Regional relationships among earthquake magnitude scale, Revs. of Geophys. and Space Phys., 19, pp. 649–663.

[23] Ritsema, R., 1981: On the assessment of seismic risk in the North Sea area, Koninklijk Nederlands Meteorologisch Instituut report, 19pp.

[24] Makropoulos, K.C., 1978: The statistics of large earthquake magnitude and an evaluation of Greek seismicity, PhD thesis, University of Edinburgh.

[25] Gumbel, E.J., 1958: Statistics of Extremes, Columbia
University Press, New York.

[26] Burton, P.W., 1979: Seismic Risk in Southern Europe
through to India examined using Gumbel's third distribution
of extreme values, Geophys. J.R. Astr. Soc., 59, pp.
249-280.

[27] Burton, P.W., 1981: Variations in seismic risk parameters
in Britain, in Proceedings, 2nd Int. Symp. Anal. Seismicity
and Seismic risk, Liblice (Czechoslovak acad. of Sciences)
Vol II, pp. 495-531.

[28] Main, I.G., and Burton, P.W., 1981: Rates of crustal
deformation inferred from seismic moment and Gumbel's third
distribution of extreme values. In Proc. Conference on
Earthquakes and Earthquake Engineering - the eastern United
States, J. Beavers, Editor; Published by Ann Arbor Science
Ltd., The Butterworth group; Vol II, pp. 937-951.

[29] Burton, P.W., 1978: Perceptible earthquakes in the United
Kingdom, Geophys.J.R.astr.Soc., 54, pp. 475-479.

[30] Lilwall, R.C., 1976: Seismicity and seismic hazard in
Britain, Inst.Geol.Sci.Seis.Bull. No. 4, HMSO.

[31] Cornell, C.A., 1968: Engineering seismic risk analysis,
Bull.Seism.Soc.Am., 68, pp. 1583-1606.

[32] Ambraseys, N.N., 1978: Preliminary analysis of European
strong motion data, 1965-1978, Bull.E.A.E.E., 4, pp. 17-37.

[33] Chiaruttini, C., and Siro, L., 1981: The correlation of
peak ground horizonal acceleration with magnitude,
distance, and seismic intensity for Friuli and Ancona,
Italy, and the Alpide Belt, Bull.Seism.Soc.Am., 71, pp.
1993-2009.

ON SOME PROBLEMS CONCERNING SEISMIC RISK ASSESSMENT FOR OFFSHORE
STRUCTURES IN THE NORTH SEA

A. Gürpınar

Specialized Engineering Consulting Services (SPECS),
Place Stéphanie 10, 1050 Brussels, Belgium

INTRODUCTION

The object of this paper is to discuss some important issues
pertaining to the seismic risk assessment of offshore structures
in the North Sea. In particular, the implications of these issues
on engineering decisions and design considerations are underlined.

A discussion of the relevant parts of the API RP 2A (1979)
is included for comparison purposes of some common problems of
assessing offshore seismicity and seismic risk. This is also
intended to bring out the comparitive risks associated with other
offshore hazards (current, wind, wave, ice) as envisaged by the
API recommendations.

The paper comprises sections on data base, modeling, low
risk problem and API Recommendations, however, the emphasis is
given to design/code aspects and the issues regarding modeling
for the assessment of seismic risks.

DATA BASE PROBLEM

The data base required for seismic risk analysis generally
comprises the following:
(a) seismic data
(b) geology/tectonics data
(c) geotechnical data

Seismic and geological data are necessary on a regional scale,
in the order of hundreds of kilometers, and are used for macro-
zoning or determination of design vibratory ground motion. Geo-

A. R. Ritsema and A. Gürpınar (eds.), Seismicity and Seismic Risk in the Offshore North Sea Area, 365–376.
Copyright © 1983 by D. Reidel Publishing Company.

technical data is required for the immediate site vicinity for
purposes of microzonation in order to delineate areas of potential
hazards such as slumping, landslides, liquefaction, sliding and
submarine volcanoes. It is assumed that the geological and geotechnical data exist in abundance for the North Sea area and it
is the seismicity data and its correlation with tectonics that is
required as future work.

Seismic data may be historical and/or instrumental both of
which generally need re-evaluation. This is a particularly difficult task for offshore areas. Historical data is at the same time
valuable and unreliable. It is often necessary to evaluate the
original accounts instead of interpreted versions of earthquake
reports. This is very cumbersome but worthwhile. Many historical
catalogs have been found to contain misleading information. This
is understandable as the primary purpose of these documents were
certainly far from establishing a basis for aseismic design of
modern structures.

Instrumental seismicity is generally used as a basis for
statistical analysis but especially in offshore areas the variation of detectability with time has to be carefully assessed.

Instrumentation may be used to obtain new recordings in
various forms and for different uses. These may vary from seabed
microearthquake instruments to strong motion monitoring of platforms. Microearthquakes have been successfully used in the identification of active structures for siting of nuclear power plants.
Some attempts to incorporate these earthquakes into the seismicity
data base, however, have been very controversial.

There is significant indication that attenuation of seismic
waves in the North Sea area is slow. This has an adverse effect
on the risk to long period structures such as offshore platforms
and pipelines. A more systematic evaluation of attenuation characteristics will help to reduce the uncertainties associated with
this parameter which in turn will be reflected in smaller standard
deviations to be incorporated in seismic risk analyses.

MODELING PROBLEMS

In order to assess the seismic risk in the North Sea area, a
model is required upon which the seismotectonic data can be incorporated. Mainly due to the strict regulations of the nuclear
industry, both deterministic and probabilistic assessments have
recently evolved into sophisticated methodologies for siting
studies of nuclear power plants. Both are based on a seismotectonic
model which identifies seismotectonic provinces or seismic sources
with characteristic parameters, i.e. maximum credible magnitudes,

frequency-magnitude relationships and attenuation relationships.

In deterministic analysis, the so called "worst" situation is investigated. The maximum credible magnitude of each of the provinces is moved to the closest proximity to the site within the province and the design ground motion parameter is calculated using the appropriate attenuation relationship. The event resulting in the maximum site value is based in this earthquake. It is understood that the maximum credible magnitudes are taken as maximum observed magnitudes, frequency-magnitude relationships are not required and the mean relationships for attenuation are utilized in deterministic assessments.

In probabilistic avaluation (Figure 1), the occurrence of earthquakes is assumed to follow a prescribed stochastic point process. Although there has been recent efforts to use Markov chains to simulate a one step memory (Padwardhan et al., 1980), the overwhelming majority of the probabilistic models are based on the Poisson assumption, i.e. statistically independent successive occurrence (Cornell, 1968; Cornell and Metz, 1973; McGuire, 1976; Gürpinar et al., 1979).

The frequency-magnitude relationships, maximum credible magnitudes and attenuation relationships are convolved into the Poisson model to yield ground motion parameters corresponding to precribed annual probabilities of exceedance. The maximum credible magnitudes may be estimated either by using fault rupture-magnitude relationships, extrapolation of frequency-magnitude curves or the use of extreme value statistics. A dispersion measure (such as the standard deviation) is incorporated in the attenuation relationship.

The obvious drawback of the deterministic approach is its intensitivity to the hazard due to sources other than the "design earthquake". The probabilistic method considers the contribution to total risk of all sources and evaluates this contribution as a function of the level of ground motion parameters.

Another important point involves the negligence of the data uncertainties when the deterministic approach is employed. Although there have been attempts to incorporate safety factors in the deterministic assessment to account for uncertainties, it is inconvenient to keep track of these quantitatively in terms of their influence on the final design values.

The probabilistic method, on the other hand, requires a fundamental understanding of the physical processes involved, in order for the results to make good sense. The choice of the reliability level is an essential decision and its limits are dictated both by the physical phenomena and cost/benefit considerations.

The amount of available and reliable data controls the input dispersion which dramatically influences the risk especially at low levels, i.e. less than 0,002 annually (or 500 years or more of average return period).

Within the context of either method, there is a need for further consideration of the site area which is generally outside of a seismotectonic province. The problem involves that of the "floating earthquake", i.e. one which is not apparently associated with a geologic feature. These events may have magnitudes up to about 5.0 (assuming that for anything bigger causative structure is identifiable) and are called near field earthquakes. They may cause very high acceleration spikes and have a high frequency content. Their duration is genrally very short, in terms of several seconds. The usual practice is to check the design (based on the seismotectonic model and deterministic/probabilistic assessment) for this near field situation. Except for possible adverse P-Δ effects, these are not considered potentially hazardous for ductile, long-period structures such as offshore platforms.

For the North Sea area, the first consideration is to address the question of seismotectonic regionalization. The fundamental issue is whether active and/or capable structures exist and may be identified. Once the basic features of such a model is established further refinements may be incorporated by long term monitoring techniques.

In the absence of a seismotectonic model or if no causative structure is identified, then the problem is one of a seismically "white" region, i.e. a region in which the probability of earth-quake occurrence in equal areas are assumed equal. This is clearly an undesirable alternative as it leads to an exercise in statistics where the data base could not be much poorer.

The issue is further complicated by the fact that there is not one "site" in consideration but a multitude of sites, clusters and network configurations. This point is significant in terms of the interests of the licensor and the licensee.

LOW RISK PROBLEM

For every physical event there are limits beyond which it is impossible to cross without shifting the scale of the envisaged phenomenon. For example a ten thousand year event is generally considered the maximum duration for the average return period of the safe shutdown earthquake for nuclear power plants. This implies an annual reccurence rate of 10^{-5} which results in an overall reliability of over 99% for the lifetime of the power plant.

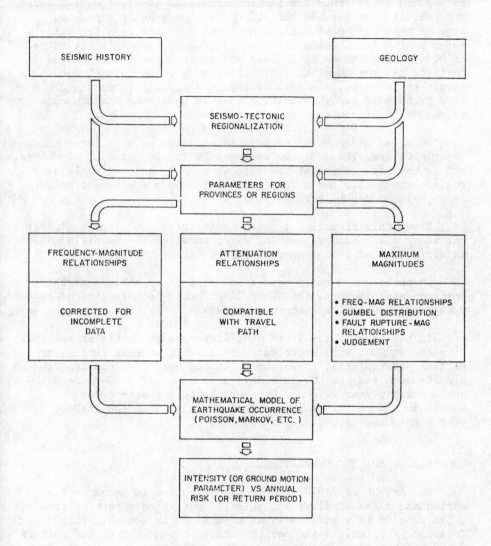

FIGURE 1
Schematic representation
of seismic risk analysis methodology
(From Gürpinar and Michalopoulos, 1980)

Depending on the physical characteristics of the source that
is assumed to generate the ten thousand year event, the risk curve
may exhibit considerable variation. For the sake of argument,
suppose that the only significant seismic source is a ten kilo-
meter active fault which is known to produce events in the range
of 4,0-5,0 magnitude. Regardless of how long the return period is
taken to be, the physical limitations of the source will not
permit an increase in magnitude beyond a certain threshold. This
is a problem of early saturation, one in which the maximum capacity
is reached in a comparatively short time.

In ocean wave analysis, this is a common problem for shallow
oceanic waters. The 100 year wave may be in the order of 30 meters,
but if the water depth at the site is only 10 meters, early saturation occurs save for a very gradual increase due to storm tide/
surge related depth increase.

Given this situation it becomes important to alter the physical model and consider breaking waves in which the particle motion
is different and mass transport may be significant.

In cases where early saturation occurs for any one of the
environmental factors, the often "improbable" combination of two
exceptional events may become critical.

In the case of the North Sea, if very high levels of reliability are sought (similar to safe shutdown earthquake levels for
nuclear power plants, for example) it may well turn out that the
simultaneous (say on the same day) occurrence of a 30-year earthquake and a 10-year wave may be probable to the same order of
magnitude and result in bigger lateral loads than the 10.000
year earhtquake alone.

COMMENTS ON API RECOMMENDATIONS

The API RP 2A (1979) Recommended Practice is based on five
earthquake zones of different seismic coefficients ranging from
0 (in zone 0) to a maximum value of 0,40 g (in zone 5). This
follows, in principle the onshore zoning of the U.S.A. but suffers
from a common flaw which has been long rectified in many other
macrozonation maps around the world. This involves the "jumps" of
ground acceleration values reflected in the API map where in fact
the transition should have been gradual. This is recognized as a
requisite for macrozonation since zonation on this scale is a
simplification of contour mapping of the design ground motion
parameter.

These jumps are located in:
. offshore Georgia from 0 to 2,
. offshore Northern California from 2 to 4,
. offshore Alaska from 1 to 3 to 5.

A second observation concerns the last two paragraphs of
Section 1.5 of the API Recommended Practice. Here, in relation
with the reliablility level for environmental loads, it is stated
that:

"As a guide, analyses have indicated that the
optimum average expected recurrence interval
is several times the planned life of the
platform".

First of all, this implies that design wind, wave and earthquake are to be computed for the same average return period.
However, the physical mechanisms causing extreme loading conditions due to these environmental phenomena possess different time
scales and the level of conservatism is therefore not consistent
when the average return period is taken equal for each. This has
already been discussed from a different angle in the consideration
of the Low Risk Problem.

Furthermore, if the lifetime of an offshore platform is taken
as, say, thirty years, the average return period for design would
be about one hundred years if the API Recommendations are followed
(Figure 2). This would correspond to a probability of exceedance
of about 30% during the lifetime of the platform which cannot be
considered as sufficiently conservative by most standards. In fact,
the earthquake zoning map of the U.S.A. (onshore) is based on a
much higher reliability level for ordinary building structures,
i.e. a probability of exceedance of 10% in fifty years corresponding
to an average return period of about 475 years.

According to the API Recommendations, earthquake loads
constitute a significant calss of environmental factors. In section
2.10 (b2), it is recommended that for zones 1 and 2 a simple analysis is sufficient to compare the seismic loads to those induced
by waves. Recalling that Zones 1 and 2 correspond to design accelerations of 5% and 10% of the acceleration of gravity respectively, and that the design wave heights are considerable both in the
Atlantic and Pacific Coasts (25-35 meters) the implication is
very informative, i.e. accelerations over 0.1 g are likely to
produce lateral loads exceeding those produced by oceanic waves.

With respect to the frequency content associated with the
design acceleration, API recommends a standardized response
spectrum, the long-period end of which is increased for "softer"
soil conditions. This spectrum is similar to the Newmark et al.
(1973) spectrum derived for nuclear power plants and which corresponds to median plus one standard deviation spectral ordinates
computed from an ensemble of actual observations.
In fact the Newmark et al. (1973) spectrum lies between the
spectra for Type B and Type C soil conditions as specified by the
API Recommendations.

Here it is important to recall that the significant part of the spectrum is characterized by the equation

$$S_A = \frac{1.8}{T} \cdot g$$

(for Soil Type C, i.e. deep strong alluvium) which governs for values of T greater than 0.7 seconds. Here T is the fundamental period of the platform in seconds, g is the acceleration due to gravity and S_A is the standardized spectral acceleration.

To have an approximate idea of the periods involved, Table 1 and Figure 3 are presented as taken from Gürpinar et al. (1980). The platform shown in Figure 3 is 117 meters high (from the seabed) with piles extending to a depth of 83 meters. From Table 1, the fundamental period of the structure is observed to be about 2,5 seconds.

This is significant from the point of view of the controling ground motion parameters. According to the results of Kobayashi and Nagahashi (1973), the base shear coefficient is best corre- lated with ground displacement for structures having fundamental periods longer than 1.5 seconds.

The corresponding period for structural deflection is about 2.5 seconds. This in turn implies that offshore structures can be classified as "displacement controlled" which required care- ful representation of the design time histories.
It should be remembered that the results may be sensitive to the base line correction techniques used in data processing (see for example Erdik and Gürpinar, 1980).

As a last remark, the way in which the API Recommendations account for the controversial issue of soil amplification will be briefly pointed out. API has followed the more recent suggestions in this regard incorporating soil effects as spectral considera- tions rather than modifications on peak acceleration values. The spectral ordinates based on over 200 records as computed by Trifunac (1980) confirm these recommendations.

CONCLUDING REMARKS

Some problem areas concerning the assessment of seismic risk in offshore areas and the North Sea in particular are outlined. A brief review of the API Recommendations is also presented in order to bring out similar aspects regarding possible North Sea applications.

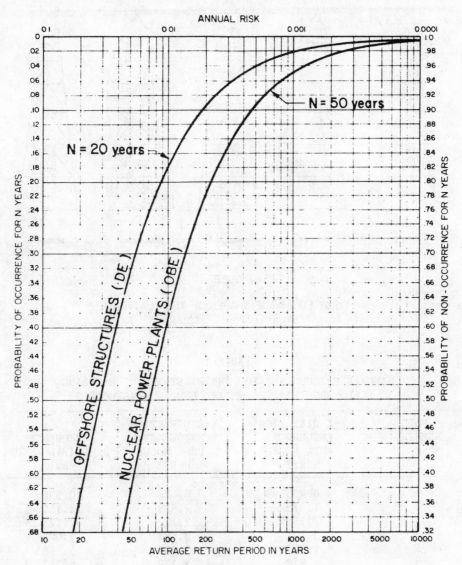

FIGURE 2
BERNOUILLI TRIAL GRAPH
(From Gürpinar and Gülkan, 1978)

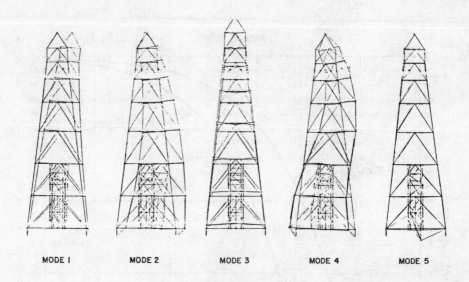

MODE I   MODE 2   MODE 3   MODE 4   MODE 5

FIGURE 3

FIRST FIVE MODE SHAPES OF THE PLATFORM
(From Gürpınar et al, 1980)

TABLE 1

NATURAL FREQUENCIES FOR THE FIRST TEN MODES USING
FULL STRUCTURAL FRAME AND LUMPED MASS MODELS

MODE NUMBER	X-Z DIRECTION FREQUENCY FULL MODEL HERTZ	Y-Z DIRECTION FREQUENCY FULL MODEL HERTZ	FREQUENCY SIMPLIFIED MODEL HERTZ
1	0.405	0.406	0.556
2	1.473	1.489	2.155
3	2.157	2.157	2.173
4	3.101	2.892	4.005
5	4.505	3.033	5.988
6	4.578	3.129	6.757
7	4.676	3.475	7.888
8	4.731	3.846	8.814
9	4.810	4.242	9.024
10	5.236	4.622	10.940

The first five mode shapes are presented in Fig. 3.
(From Gürpınar et al., 1980)

REFERENCES

American Petroleum Institute (API), 1979. Recommended Practice
    for Planning, Designing and Construction of Offshore Plat-
    forms, API RP 2A, Washington D.C.

Cornell, C.A., 1968. Engineering Seismic Risk Analysis, Bulletin
    of the Seismological Society of America, vol.58, No. 5.

Cornell, C.A. and H.A. Merz, 1974. Seismic Risk Analysis of Boston,
    Journal of Structural Division, ASCE, vol. 101, No. ST10.

Erdik, M and A. Gürpinar, 1980. The Strong Motion Program and
    Associated Developments in Turkey, Proc. of the Research
    Conference on Earthquake Engineering, Skopje, Yugoslavia.

Gürpinar, A. and P. Gülkan, 1977. Preliminary Seismic Risk Assess-
    ment for Siting of Nuclear Power Plants, Proc. of the Sym-
    posium on the Analysis of Seismicity and on Seismic Risk,
    European Seismological Commission, Liblice.

Gürpinar, A., M. Erdik, S. Yücemen and M. Öner, 1979. Seismic Risk
    Analysis of Northern Anatolia Based on Intensity Attenuation,
    Proc. of the Second U.S. Conference on Earthquake Engineering,
    Stanford University.

Gürpinar, A. and A.P. Michalopoulos, 1980. Seismic Risk Assessment
    of Nuclear Power Plants for Siting and Design, Proc. of the
    CSNI Specialist Meeting on Probabilistic Methods for Seismic
    Risk Assessment of Nuclear Power Plants, OECD-NEA, Lisbon.

Gürpinar, A, B.J. Gryspeert and J.C. Baker, 1980. Dynamic Analysis
    of Offshore Platforms under Seismic Excitation, Proc. of the
    Seventh World Conference on Earthquake Engineering, Istanbul.

Kobayashi, H. and S. Nagahashi, 1973. Intensity of Earthquake
    Motion for the Design of Structures, Tokyo Institute of
    Technology.

McGuire, R.K., 1976. FORTRAN Computer Program for Seismic Risk
    Analysis, USGS Open File Report 76-67.

Newmark, N.M., J.A. Blume and K.K. Kapur, 1973. Design Response
    Spectra for Nuclear Power Plants, Proc. of the ASCE, Struc-
    tural Engineering Meeting, San Francisco.

Padwardhan, A.S., R.B. Kulkarni and D. Tocher, 1980. A Semi-Markov
    Model for Characterizing Recurrence of Great Earthquakes,
    Bulletin of the Seismological Society of America, vol. 70,
    No. 1.

Trifunac, M.D., 1980. Effects of Site Geology and Amplitudes of
    Strong Earthquake Ground Motion, Proc. of the Seventh World
    Conference on Earthquake Engineering, Istanbul.

UNCERTAINTY IN U.K. SEISMIC HAZARD ANALYSIS

Gordon Woo

Principia Mechanica Limited - London

Uncertainty in seismic hazard analysis arises out of the intrinsic variability of seismological parameters and also the dearth
of knowledge and understanding of earthquake activity. For regions
of the world such as the United Kingdom where significant earthquakes occur infrequently, the prospects for the recording of new
important data in the near future are slim, and the maximum possible use of existing data must be made if the uncertainties associated with ignorance are to be reduced. Some ways in which this
can be achieved are outlined here, in an argument in favour of
lowering uncertainty through further research than accepting a much
higher level of statistical error as a necessity.

In those regions of the world which are recognized to be particularly seismic and where significant earthquakes occur at a rate
sufficient to warrant the installation of strong-motion instrumental networks, local data can be accumulated within a comparatively brief period to allow an informed engineering seismic hazard
analysis to be performed.

In contrast are those other regions -generally intraplate,
where the return period for major events us so high as to render
large-scale instrumentation schemes impractical and where the continuing absence of adequate strong-motion data will hamper attempts
at local hazard analysis for the foreseeable future.

Great Britain and its surrounding waters is in the latter
category, as that the analyst charged with the task of hazard assessment for this region has to be resigned to the comparative poverty
of local contemporary instrumental data. Confronted with this challenge, three possible avenues of approach exist :

A. R. Ritsema and A. Gürpınar (eds.), Seismicity and Seismic Risk in the Offshore North Sea Area, 377–383.
Copyright © 1983 by D. Reidel Publishing Company.

(a) The problem can be acknowledged to be intractable and a solution sought through the solicitation of expert opinion and the application of Bayesian statistics ;

(b) The problem can be tackled to the fullest extent possible using all available sources of information : historical, geological, tectonic and instrumental ;

(c) The problem can be addressed in succession through (b) and then (a).

The quantification of uncertainty is an important element in hazard analysis, and fairly small differences in distribution parameters of the main seismological variables can have a marked effect on the final results. However it is all too common for the distinction between intrinsic stochastic variability and statistical error associated with ignorance or mis-information to be blurred ; it is far easier to broaden variances under the pretext of comprehensiveness than to reduce them through a programme of seismological investigation.

If there is a place for the introduction of Bayesian techniques of analysis, it must surely come only after every possible means of objective earthquake research has been followed up. Through perseverence with seismic studies, many mistakes can be corrected and coefficients of variation reduced. A few examples taken from the Principia re-assessment of British seismicity will be quoted here as illustration.

Consider first the issue of the focal depths at which earthquakes originate within the British Isles. Accurate instrumental determinations of focal depths require data from seismograph arrays, little of which exists for significant British events, although a substantial set of LOWNET readings for minor events has been collected over the past twenty years by the IGS .

In the absence of instrumental data for past important British earthquakes, macroseismic information has previously been used as a basis for depth estimates. While most of these estimates have tended to indicate shallow depths, a few, notably some concerning events in Herefordshire, have suggested depths of up to 60 kms (Lilwall 1976). This is somewhat paradoxical given the thin crustal structure of these islands and the posited plastic state of the base of the lithosphere and the asthenosphere.

The resolution of this paradox is in the re-examination of the primary intensity data upon which the isoseismal maps of these supposedly deep events have been drawn. The return to original sources of information reveals a much sharper fall-off of intensity in the fault-zone than hitherto recognized and that a figure closer to 20 km rather than 60 km would be supported by the revised intensity analysis.

Thus while in previous UK seismic hazard analyses, focal depth distributions have been smeared out to allow for the hypothetically deep events, in fact the true distribution is much more clearly defined and consistent with geophysical expectations.

The point made here is not merely a technical one on focal depths, but is more general : uncertainty can be reduced by systematic programmes of research. Expert opinion can only be as good as the best knowledge and information at hand, and diversity in such opinion is a reflection of the need for further research, not more intricate statistical analysis. The enormous variety and differences in opinion expressed is an argument not for calling in a new team of experts or for upgrading it or enlarging it, but is an argument for more seismological studies of the region.

Seismological studies have traditionally followed a fairly standard pattern involving macroseismic and magnitude work, but little in the way of actual seismogram frequency analysis. This is unfortunate since the latter can contribute greatly to the control of uncertainties arising from intensity assignments. The fact that this is possible is due to several correlations which exist between intrumental and MMI parameters. These correlations assist in unifying the qualitative with the quantitative seismological parameters which otherwise would appear disjoint.

One relation of this kind is that which links M_s with A_{III} or A_{IV}, the areas within isoseismals III and IV. Not only does this provide a numerically very stable method of computing magnitudes since the values calculated using A_{III} and A_{IV} differ no more than by 0.1 or 0.2, but it also provides an indirect check on the areas themselves. This is particularly useful for validating intensity assignments in situations where intensity is ill-defined such as in offshore regions or in areas of low population density.

Another relation is one which links seismic moment M_0 with the higher isoseismal A_{VI}. This also provides an indirect check on intensity assignments and thereby reduces the uncertainty in the manner and method used to obtain them. To serve as an example of this, the North Sea event of June 7th 1931 seems particularly appropriate for this meeting, which is centred on North Sea seismicity.

From seismograms recorded at Hohenheim and Copenhagen, a value of M_0 of 10^{24} dyne-cm has been calculated by Seeber, Armbruster and Mori of the Lamont Geological Observatory. For a moment of this moderate size, a plot of Hermann et al (1978) which is reproduced

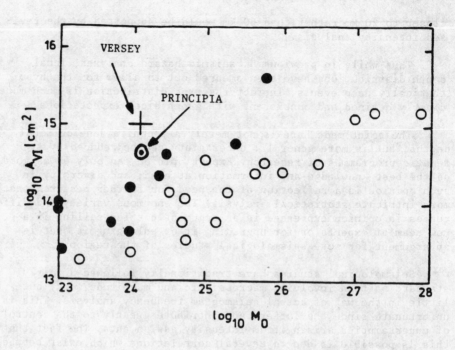

<u>Fig. 1</u> : PLOT OF M_O AND A_{III}.

BLACK DOTS INDICATE EASTERN US EVENTS -
OPEN DOTS INDICATE CALIFORNIAN EVENTS
(AFTER HERMANN 1978)

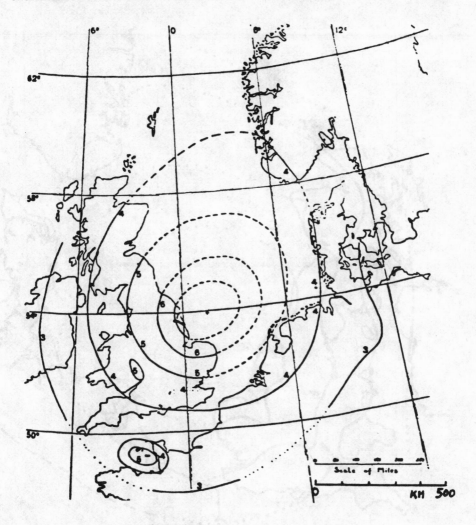

Fig. 2 : INTENSITY MAP OF JUNE 7 1931
          (AFTER VERSEY 1939)

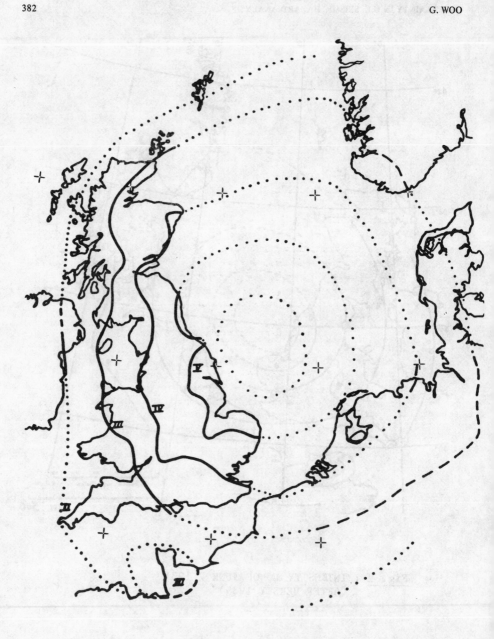

Fig. 3 : PRINCIPIA INTENSITY MAP OF JUNE 7 1931 EVENT

as Fig. 1, shows that a felt area of 100,000 kms for A_{VI} would be unusually high. Yet Versey (1939), in his study of the earthquake, would claim a value of 110,000 km for A_{VI}, based on his intensity map shown in Fig. 2. This is more than twice the area for A_{VI} newly estimated by Principia from a completely revised map (Fig. 3) drawn from original data sources. This much lower figure tallies much better with the moment of the event than the earlier figure.

This illustration is a paradigm of the effectiveness of research effort in uncovering errors and reducing uncertainties. Naturally the gaps in instrumental (especially strong-motion) coverage in Britain cannot be made good by research alone, but many grey areas of ignorance can be clarified through extensive programmes of seismological study. In this way many sources of statistical variability can be eradicated to the improvement of probabilistic seismic hazard analysis.

References :

Hermann R.B., S. H. Cheng and O. Nuttli (1978)
"Archaeoseismology : The New Madrid earthquake of 1811-1812"
Bull. Seism. Soc. Amer. 68.   pp. 1751-1759.

Lilwall R.C. (1976) "Seismicity and Seismic Hazard in Britain"
IGS Seism. Bull No. 4.

Principia (1982) "British Earthquakes" Report to CEGB, SSEB, BNFL (UK). June.

Seber L., J.G. Armbruster and J. Mori (1982)
"Spectral Characteristics and Size of the June 7th 1931 North Sea Earthquake" Lamont Geological Observatory Report. May.

Versey H.C. (1939) "The North Sea Earthquake of June 7th 1931"
Monthly Notices of the Royal Astronomical Society.   pp. 416-423.

METHODOLOGICAL ASPECTS ENCOUNTERED IN THE LOWER RHINE AREA
SEISMIC HAZARD ANALYSIS

W. Rosenhauer

INTERATOM, 5060 Bergisch Gladbach
Federal Republic of Germany

A generalized extreme value distribution is presented as a
tool for the magnitude characterization of seismic source
areas, including parameter estimation formulae and an application. Concerning the source to site transfer the concept
of individual intensity attenuation laws and its realization by simulated macroseismic load collectives is described.
The load collectives exhibit the most probable ranges of
focal parameters consistent with a given site intensity.
Response spectra have consequently been derived from strong
motion records belonging to appropriate focal and soil conditions, avoiding the usual employment of peak acceleration and
its poor correlation to intensity.

1. INTRODUCTION

Design against external events has been an important subject
for construction and licensing of the Kalkar fast breeder
prototype reactor SNR-300 and has already required an
engineering in the order of several hundred man-years.
This considerable effort was accompanied by seismic site
analyses, e. g. / 1 /, which were performed in cooperation
with Erdbebenstation Bensberg, Universität Köln.

The ground motions adopted for nuclear plant design are
much stronger than measured time histories available for the
site region, just because - among a lot of other reasons - the
design earthquakes have to be placed well above the events experienced in the past. The inevitable extrapolations, conser-

A. R. Ritsema and A. Gürpınar (eds.), Seismicity and Seismic Risk in the Offshore North Sea Area, 385–396.
Copyright © 1983 by D. Reidel Publishing Company.

vative assumptions, uncertainties etc. are best treated in a
valuing and transparent manner by a probabilistic model,
covering the stochastic nature of future earthquakes. The
decisive and indispensable contingent is of course the general
and local geological and seismological information (see the
contribution of L. Ahorner, Universität Köln to this workshop)
which is reflected and summarized by the model. In this paper
results with respect to the probabilistic methods are presented,
which might also be applied to the special field of interest of
the workshop, the offshore North Sea area.

2. GENERALIZED GUMBEL DISTRIBUTION

The most important site specific input laws for a seismic hazard analysis are the occurrence rates $\lambda(>M)$ specifying the
frequency of earthquakes exceeding given magnitudes M in the
surrounding seismic source areas. Various shapes for the curve
$\lambda(>M)$ have been assumed. The determination of $\lambda(>M)$ via the
probability of extremes by

$$\lambda(>M).T = - \ln P(\text{all magnitudes} \leq M \text{ in a period } T) \qquad (1)$$

seems to be the least ambiguous approach, since P(M) is known
to be in good approximation either a type I or a type III asymptotic extreme value distribution. It is convenient to use the
following formulae comprising both types:

P(all magnitudes $\leq$ M in T)=

$$G_v(M) = \exp \left[-(f_1 - f_2 \frac{M-m}{\sigma})^{1/\tau} \right], \quad 0 \leq \tau \leq 1, \quad \sigma > 0$$

$$f_1 = f_1(\tau) = \Gamma(1+\tau) \quad f_2 = f_2(\tau) = \sqrt{\Gamma(1+2\tau) - f_1^2} \qquad (2)$$

$$\Gamma(x) = \int_0^\infty t^{x-1} \exp(-t) dt \quad (\text{Gamma-Function})$$

$$M \leq M_{max} = m + \sigma \, f_1/f_2 \quad (\text{Upper magnitude bound})$$

The complexity of (2) is compensated for by very simple estimates for its parameters, which are essentially the moments of
the distribution:

$$m = < M > \cong M1 \quad (\text{mean value})$$

$$\sigma = \sqrt{<(M-m)^2>} \cong \sqrt{\frac{N}{N-1} \frac{\sqrt{M2 - M1^2}}{1-[0.1+2(0.4-\tau_s)^2]/N}} \qquad \begin{array}{l}(\text{standard} \\ \text{deviation})\end{array}$$

$$r = \left\langle \left(\frac{M-m}{\sigma} \right)^3 \right\rangle$$

$$\cong r_s = \frac{\sqrt{N(N-1)}}{N-2} \frac{M3 - 3 \cdot M2 \cdot M1 + 2M1^3}{(M2 - M1^2)^{3/2}} \quad (1 + 6/N) \quad (\text{skewness}) \qquad (3)$$

$$\tau \cong \tau_s = \begin{cases} 0 & (r_s > 0.88) \\ 0.282 - 0.317 r_s & (-0.69 \leq r_s \leq 0.88) \\ > 0.5 & (r_s < -0.69) \end{cases}$$

M1, M2, M3 are the unshifted empirical moments calculated from extremes M_1, M_2, ..., M_N observed in N periods T:

$$Mk = \frac{1}{N} \sum_{i=1}^{N} M_i^k \qquad (k = 1, 2, 3) \qquad (4)$$

The estimates (3) are unbiased, thus avoiding the errors revealed by Knopoff and Kagan / 2 /.

Nr. i	Period	M_i	Nr. i	Period	M_i
1	1750–59	(4.1)	12	1860–69	4.7
2	1760–69	4.0	13	1870–79	4.8
3	1770–79	(3.6)	14	1880–89	4.5
4	1780–89	4.8	15	1890–99	3.5
5	1790–99	3.0	16	1900–09	4.3
6	1800–09	(4.0)	17	1910–19	4.6
7	1810–19	(3.8)	18	1920–29	4.4
8	1820–29	5.4	19	1930–39	5.9
9	1830–39	(3.9)	20	1040–49	5.0
10	1840–49	5.4	21	1950–59	4.9
11	1850–59	4.2	22	1960–69	4.6

Table 1: Extremal local magnitudes M_L for the Renish and Belgian earthquake zone (except the western part of the Lower Rhinegraben, zones 2, 3, 4 in / 1 /) A = $3.81 \cdot 10^4$ km², T = 10 years, N = 22, estimated values in parentheses, compiled by L. Ahorner, 1974

	m	σ	τ	M_{max}
Estimate according to (3)	4.427 ± 0.146	0.686	0.243 (0.05 -0.40)	6.9 (≥ 6, $<\infty$)
Maximum Likelihood estimate	4.427	0.666	0.263 (0.08 -0.43)	6.7 (≥ 5.9, $<\infty$)

Table 2: Parameters of the generalized Gumbel distribution estimated from the extremal magnitudes in Table 1. (M1 = 4.427 M2 = 20.047 M3 = 92.736 r_s = 0.124)

$G_Y(M)$ becomes identical to the original Gumbel distribution $G(M)$ (type I, see e. g. / 1 /) for $\tau=0$ ($M_{max} = \infty$), whereas an estimate $M_{max} < \infty$ is obtained if $\tau_s > 0$ is significantly established. This can be checked using the τ_s uncertainty. The maximum likelihood method (corrected for bias) was determined to be more exact than the estimate (3) with

$$\tau \cong \tau_s \pm \Delta\tau_s \qquad \Delta\tau_s = 1.4 \cdot N^{-2/3} \quad (N \geq 10) \qquad (5)$$

The main advantage of the distribution (2) is that the essential part of the curve is almost independent of τ or the exact value of M_{max}, parameters which are therefore not required with high accurracy. Several applications resulted in values of τ near to 0.25 and corresponding estimates

$$M_{max} \cong m+(3 \ldots 4.5)\sigma \qquad (6)$$

in good agreement with deterministic assessments of the upper magnitude bound, suggesting that $\tau \cong 0.25$ might be an acceptable general assumption if a more exact determination is not possible.

An example illustrating the method is given in tables 1 and 2. Fig. 1 shows $\lambda(>M)$ as it was finally adopted using $M_{max} \cong 6.5$ coincident with / 1 /, resulting in insignificant changes of m, σ and τ of table 2.

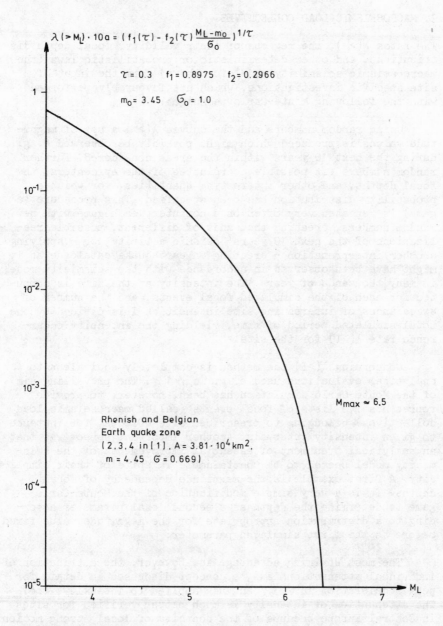

$$\lambda(>M_L) \cdot 10a = \left(f_1(\tau) - f_2(\tau)\frac{M_L - m_0}{\sigma_0}\right)^{1/\tau}$$

$\tau = 0.3 \quad f_1 = 0.8975 \quad f_2 = 0.2966$

$m_0 = 3.45 \quad \sigma_0 = 1.0$

$M_{max} \approx 6.5$

Rhenish and Belgian
Earth quake zone
$(2,3,4 \text{ in }[1]), A = 3.81 \cdot 10^4 \text{ km}^2,$
$m = 4.45 \quad \sigma = 0.669)$

Generalized Gumbel distribution
for magnitudes M_L, normalized to
$A_o = 10^4 \text{ km}^2, \quad T = 10a$

Fig. 1

3. MACROSEISMIC LOAD COLLECTIVES

The rates $\lambda(>M)$, the regions of their validity, focal depth distributions, and other deterministic or probabilistic laws (the
macro-seismic seismicity model for short) form the input of
site specific investigations, which are favourably performed
with the following Monte-Carlo technique.

Using random numbers and the curves $\lambda(>M)$ a set of magnitude values is produced which might possibly be observed e. g.
during the next 10 years within the areas considered. Further
random numbers fix possible coordinates of the epicenters, the
focal depths, and other interesting quantities, for which a
probability distribution has been specified. This procedure (a
game) is repeated very often on a computer, each game with new
random numbers, creating thousands of different, possible rea-
lizations of the next 10 years' seismic activity, or - applying
another interpretation - creating an earthquake catalogue as it
might have been observed in accordance with the seismicity model
in many thousands of years. The intensity at the site is computed for each of the simulated focal events, and the number of
exceedances of interesting site intensities I is divided by the
total simulated period of time, yielding the intensity occurrence rate $\lambda(>I)$ for the site.

Concerning $\lambda(>I)$ the method is completely equivalent to
analytical evaluations used e. g. in / 1 /. The prevailing aim
of the Monte-Carlo simulation has been, however, to provide
representative lists of focal events (called macroseismic load
collectives) belonging to prescribed site effects, for instance
to given intensity intervals. Another important aspect was that
an analytical treatment of essential improvements of the seismicity model seemed to be unattainable in spite of their simplicity. A first example is the magnitude dependency of focal
depths. It is a very simple modification of the Monte-Carlo
game to determine the depth as a second focal parameter according to a distribution appropriate for the magnitude value fixed
before as the first simulated parameter.

The most striking advantage is, however, the attribution of
individual attenuation laws, a concept discussed in detail for
peak acceleration in / 3 / and now applied to intensity, too.
The attenuation of intensity is much better verified for sites
in Central Europe because of the scarcity of local strong motion
measurements. A specific magnitude and hypocentral distance dependency of the intensity was derived from local events by
L. Ahorner (see his contribution). Included was a quantification
of possible deviations from the mean curve, ascertaining the
generally known fact that earthquakes with the same magnitude

and focal distance could lead to site intensities about one
degree lower or higher than the expected value. The interpretation of results derived with a mean curve is difficult, for
that reason. The Monte-Carlo simulation allows to supply each
of the simulated focal events with its individual attenuation
curve, thus taking into account that - varying from earthquake
to earthquake - the attenuation might be much slower or faster
than usual. This procedure leads in general to more conservative
occurrence rates with significantly reduced error bands, / 3 /,
since a main contribution to the uncertainties of results calculated with the mean attenuation is incorporated in the simulated
probability itself.

Lots of macroseismic collectives have been simulated for
Kalkar/SNR-300, based on the seismic source areas and original
data of / 1 /. Fig. 2 shows an example displaying the magnitude
range of earthquakes with site intensities $I = 7.5 \pm 0.25$ and
hypocentral distances $R = (25 \pm 10)$ km, as it is resulting from
competing magnitude occurrence rates, attenuation curve uncertainties, and other influences mentioned above.

4. SITE SPECIFIC RESPONSE SPECTRA

The macroseismic load collectives offer various approaches to
the definition of site specific seismic loads in terms of
engineering quantities, as required for design. The bulk of
relevant seismicity observations has been described by intensity. Starting from a site intensity at an adequately low
probability level and defining acceleration response spectra
calculated from time histories measured elsewhere was chosen
as a method very similar to the access of the licensing procedure.

The usually applied standardized spectra cannot, however, be
expected to characterize the seismic reality at the site, due
to their generation. They are afflicted, moreover, with peak
acceleration as a scaling factor. The high frequency portion
of the spectra, i. e. the peak acceleration, is in general not
significant for nuclear plants. It cannot be correlated in a
satisfactory manner to intensity even for more active regions
of the world, perhaps because the intensity classification is
possibly accounting for low frequency effects above all. Direct
correlations of the required low frequency domain (≤ 20 Hz) to
intensity are not in use and should meet difficulties because
of their variation with the extremely different focal distances,
magnitudes etc. all consistent with a given intensity value.

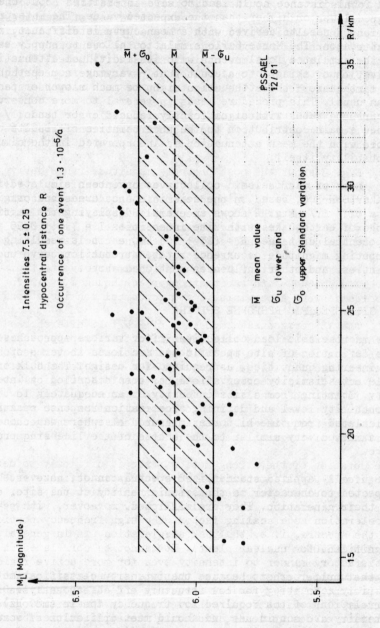

Intensities 7.5 ± 0.25
Hypocentral distance R
Occurrence of one event 1.3 · 10⁻⁶/a

$\overline{M}$ mean value
$\overline{G}_u$ lower and
$\overline{G}_o$ upper Standard variation

PSSAEL
12/81

$\overline{M} + G_o$ $\overline{M}$ $\overline{M} - G_u$

R/km

M_L(Magnitude)

Macroseismic load collective
for Kalkar

Fig. 2

This is the gap where the macroseismic load collectives step in. They display by a representative set of examples the variety and relative frequency of focal parameters belonging to a site intensity interval according to the occurrence probabilities implied by the macroseismic seismicity model. This identification of typical focal values helps to make a proper choice of time histories for the computation of mean response spectra.

As an additional limitation for site specific spectra only records for comparable soil and subsoil conditions should be selected.

Fig. 3 shows the most probable magnitude range derived from 3 macroseismic load collectives for intensities $6.5 \pm 1/4$, $7 \pm 1/4$, and $7.5 \pm 1/4$ around the safe shutdown earthquake $(I = VII)$ of the SNR-300. Hypocentral distances were limited to $R = (25 \pm 10)$ km, being the range most often met (≥ 50 %) in the macroseismic load collectives. One conclusion of the probabilistic analysis exceeding naked probability figures was that earthquakes with $M \cong 6$ and $R \cong 25$ km are events typical for the safe shutdown and a bit stronger earthquakes, well in accordance with deterministic geotectonical considerations.

The 12 horizontal components of 3 Friuli aftershocks recorded at two stations were identified to be representative events consistent with the criteria mentioned above. Response spectra (see fig. 4) were computed as a basis for the definition of a design spectrum. The important methodological aspect is that no normalization at high frequencies has to be performed before averaging, and no renormalization with peak acceleration is required as a consequence, and that the genuine uncertainty of the mean spectrum is obtained for all frequencies.

The duration of the strong motion phase is difficult to determine for a given intensity. A more realistic estimate (see fig. 4) resulted from the analysis as an additional important input for inelastic structural analyses.

5. CONCLUSIONS

A great number of stochastic, unknown, or uncertain parameters is present in the dynamical structure and safety analyses of nuclear plant vibrational behaviour and in the seismological definition of design loads, excluding the application of worst-case assumptions. Detailed and realistic descriptions of the ground motions induced at plant sites by future earthquakes are required in order to guarantee that seismic design and safety

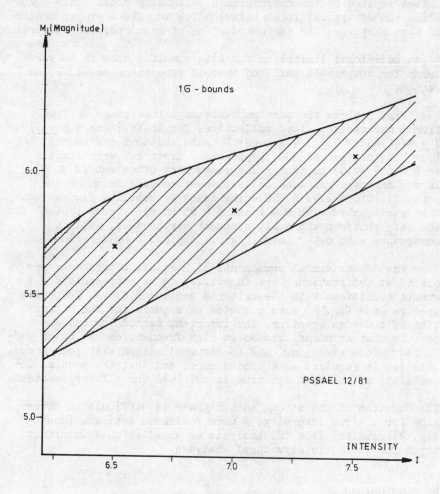

Magnitudes for intensities I ±0,25,
hypocentral distances R=(25±10)km   Fig. 3
Site: Kalkar

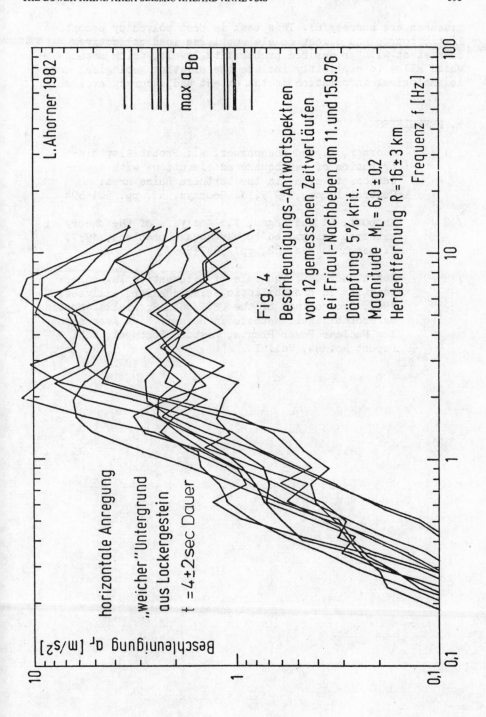

Fig. 4

Beschleunigungs-Antwortspektren
von 12 gemessenen Zeitverläufen
bei Friaul-Nachbeben am 11. und 15.9.76
Dämpfung 5% krit.
Magnitude $M_L = 6,0 \pm 0,2$
Herdentfernung $R = 16 \pm 3$ km

L. Ahorner 1982

$\max a_{B0}$

horizontale Anregung
„weicher" Untergrund
aus Lockergestein
$t = 4 \pm 2$ sec Dauer

Beschleunigung a_r [m/s²]

Frequenz f [Hz]

measures are successful. This task is best solved by probabilistic source and effect models supplying load collectives
consistent with prescribed probability levels, using methods
which allow to explicitly include the existing geological and
seismological information in the extent and accuracy available.

5. REFERENCES

/ 1 /   Ahorner, L. and Rosenhauer, W.: Probability dis-
           tribution of earthquake accelerations with appli-
           cations to sites in the Northern Rhine area,
           Central Europe, 1975, J. Geophys. 41, pp. 581-594

/ 2 /   Knopoff, L. and Kagan, Y.: Analysis of the theory of
           extremes as applied to earthquake problems, 1977,
           J. Geophys. Res. 82, pp. 5647-5657

/ 3 /   Rosenhauer, W.: The role of attenuation law uncer-
           tainties in probabilistic seismic hazard analyses,
           1980, Proceedings of the OECD/CSNI Specialist Meeting
           on Probabilistic Methods in Seismic Risk Assessment
           for Nuclear Power Plants, Lisbon, Portugal, CSNI
           Report No. 44, Vol. I

OUTLINE OF PROCEDURE FOR TOTAL SEISMIC RISK ESTIMATION

O.A. Sandvin

NTNF/NORSAR, P.O. Box 51, N-2007 Kjeller, Norway

To demonstrate a systematic approach for establishing estimates of total seismic risk as associated with some well-defined
geographical area and its objects, I will give a brief outline
of the definitions and theoretical aspects as adapted by Keilis-
Borok et al. (1974) at the Institute of Physics of the Earth in
Moscow in their approach to the problem. The systematic incorpo-
ration of all pertinent information bearing on the total seismic
risk has resulted in a complex mathematical framework and a widely
applicable model characterized by its generality. A computer
version of the model is also available and is quite simple to
adapt in a practical application as long as sufficient and reliable
data is present. As a curiosity it might be mentioned that this
approach has been adapted for the estimation of seismic risk
associated with the construction of the Trans-Siberian Railway
(BAM-project).

However, as has been clearly demonstrated as one of the main
conclusions at the Workshop, the lack of consistent data pertaining
to the problem of risk estimation connected to the North Sea area
is the major obstacle in using a sophisticated mathematical model
with some degree of reliability. But in due course of the upgraded
activity in data collection and data preparation tied to the oil
exploration business within the area combined with the interest
of relevant international scientific societies in increasing the
data potential, as this Workshop is an indication of, this might
result in a significant improvement in data quantity as well as
in quality for future research. An important aspect of this
Workshop is to integrate all the different scientific disciplines
having some relevance to the seismic risk estimation of the North
Sea and to point out the basic ideas for future recommendations

A. R. Ritsema and A. Gürpınar (eds.), Seismicity and Seismic Risk in the Offshore North Sea Area, 397–404.
Copyright © 1983 by D. Reidel Publishing Company.

in a joint cooperation to optimize the activity and improve the
relevant information potential. It is within this context that
the present risk estimation model may prove its potential and
usefulness. Due to its generality and flexibility this computer
program opens for the inclusion of a lot of different parameters
pertaining to the overall problem of risk estimation on an interdisciplinary level.

Besides estimating the seismic risk in terms of probability
level of ground acceleration exceeding certain levels at certain
locations within a well-defined period of time, as is the case
for competing computer programs estimating seismic risk levels,
this particular program also has the potential of estimating the
risk in a much wider sense. In defining a number of different
objects within the area of interest one is left with the possibility of defining various effects connected to the objects as a
function of ground acceleration and the computer program will
estimate the expected values of your effects and the distribution
function within a probalistic framework. In addition the model
is dynamic in the sense that information may be supplied to the
computer program describing the functional relationship between
the various effects and time and the computer program may subsequently give you an estimate of the dynamic behaviour of the
cost of the effect as a function of time.

In order to obtain realistic estimates from the program more
information must be supplied and in particular information from
various disciplines, i.e., to specify the effect of an oil-produ-
cing platform in the North Sea there is a substantial need for
additional information describing the dynamic response of the
platform, soil or coupling effects as well as an economic consequence analysis of the various levels of effects that might occur.

The mathematical modelling is based on the following knowledge:
- the sequence of earthquakes in time, space and their energy
  distribution;
- spatial distribution of intensity of ground tremors from fixed
  earthquakes;
- effects that given object experience from ground tremors of a
  certain intensity.

The sequence of earthquakes is treated as a stochastic process as is the case in other computer programs estimating risk;
the probability distribution frunction is supposed to follow a
Poisson-distribution, but still with the possibility to implement
your own distribution function.

The two first points above reflect the information needed for
estimating the probability function for ground acceleration
exceeding defined values while the third point introduces some

kind of consequence analysis based on additional information.

The computer version of the model permits the use of all pertinent information, such as
- Engineering
- Seismological
- Economic.

As examples of effects might be:
- tremors of an object: the total size of those parts of the objects which experience tremors of a given intensity;
- total damage of the object from seismic tremors;
- size of population or other valuable commodities present in the zones of tremors of given intensities.

Some of the definitions used in the model are stated below for a proper interpretation of the formulas to appear later.

The earthquake sequence is regarded as a random event (t,g,e) where
t = actual time interval
g = vector coordinates of the epicenter within the seismic region
e = measure of energy (magnitude, intensity).

Within the actual geographical area under consideration two different subsets are distinguished
- object 0 (a restricted area where the seismic risk is estimated)
- danger zone g' (seismic active area).

The effect $x(t,g,e)$ is associated with each earthquake and characterizes the damage to the object caused by earthquake (t,g,e).

The problem is that of calculating the statistical distribution function $F(x)$ for the total effect

$$X_\Sigma(t) = \sum_{0 < t_i < t} x(t_i, g_i, e_i)$$

caused by the sequence of events (t_i, g_i, e_i).

Some of the hypotheses of the models are:
- Model of the earthquake sequence (t,g,e) follows a Poisson flow in the volume txgxe, i.e., number of events in nonoverlapping volumes $\Delta v = \Delta t \Delta g \Delta e$ are statistically independent with distribution function

$$P(N(\Delta v)) = \frac{\lambda^k(\Delta v)}{K!} \exp(-\lambda(\Delta v))$$

with parameter

$$\lambda\ (\Delta v)\ =\ \lambda\ \Delta t\ \underset{\Delta g\Delta e}{\iint}\ P(g)F(e/g)dgde$$

- Effects $x(t,g,e)$ from various fixed earthquakes are statistically independent.

- Earthquakes (t,g,e) cause at each point $\widetilde{g} \in G$ tremors of which the intensity is defined by the random function $I(\widetilde{g}/g,e)$, i.e., tremors are of intensity c and higher in the random area $\sigma(c) = \{\widetilde{g}:I(\widetilde{g}/g,e) \geqslant c\}$

- The effect of earthquake $x(t,g,e)$ is the sum of the effects caused by tremors at each point of the object.

- A tremor of intensity c causes at each point of an object $g \in 0$ a determined effect, defined by the function

$$x(t,g,e) = P(t,g)\ x_0(g,c)$$

Here $x_0(g,c)$ is the effect of the tremor (g,c) at an initial moment in time and $P(t,g)$ is a nonrandom function describing the dynamics as function of time.

SOLUTION OF THE PROBLEM

Under the above hypothesis the distribution function $F^{\Sigma}(x)$ of the total effect is a generalized Poisson distribution:

$$F^{\Sigma}(x)\ =\ \sum_{n=0}^{\infty}\ \frac{(\lambda t)^n}{n!}\ \exp\ (-\lambda t)F^{(n)}(x)$$

where $F^{(n)}(x)$ is the n-th convolution of $F(x)$.

$$F(x)\ =\ \frac{1}{T}\ \underset{TGE}{\iiint}\ F(x/t,g,e)P(g,t)f(e/g)dtdgde$$

$F(x)$ is the distribution of the effect of a single arbitrary earthquake in txgxe and $f(e/g)\Delta e$ is the conditional probability of the appearance of an event with energy $e \in \Delta e$ on the condition that it appeared at point g.

ANALYTIC PRESENTATION OF I_O

The isolines $I_O = c$ of function $I_O(\widetilde{g}/g,e)$ are assumed to be concentric ellipses with center at g. The ellipses are defined by three parameters: area Q, ratio of axis L and azimuth A of long axis. Area $Q_c(e,g)$ inside the isoseismals are represented by the model

$$\lg Q_c(e/g) = \lg \hat{Q}_c(e/g) + \Delta Q(g)$$

where $\lg \hat{Q}_c$ defines the mean value of the logarithm of area $Q_c(e/g)$ and $\Delta Q(g)$ is a random addition.

MODEL OF ISOSEISMALS

The intensity of tremor I at point g from an earthquake (g,e) is defined by:

$$I(\widetilde{g}/g,e) = I_O(\widetilde{g}/g,e) + \Delta I(\widetilde{g})$$

where $I_O(\widetilde{g}/g,e)$ corresponds to the intensity of tremor at point g for areas with normal soil.

$\Delta I(g)$ is the determined correction for I_O estimated from local soil conditions.

The orientation of isoseismal areas is defined by the azimuth A of the largest axis. It is assumed that

$$A = \bar{A}(g) + \Delta A(g,e)$$

where $\bar{A}(g)$ is average direction of extension of isoseismal area and $\Delta A(g,e)$ is random correction for average azimuth, which is defined by the distribution function $F_{\Delta A}(./g,e)$.

The ratio of the major axis of an ellptical isoseismal is a non-random function

$$L(e,g) \geqslant 1$$

The connection between $\lg Q_c$ and magnitudes is approximately linear within some magnitude intervals:

$$\ell g \hat{Q}_c(m/g) \;=\; A_c(g) + B_c(g)m \qquad\qquad M \geqslant M_c$$

$$\hat{Q}_c(m/g) \;=\; 0 \qquad\qquad\qquad\qquad M < M_c$$

The distribution function of correction ΔQ is assumed to be of the form

$$F_{\Delta Q}(x/g) \;=\; \frac{\emptyset(s)-\emptyset(-K)}{1-2\emptyset(-K)} \qquad\qquad |x| < KS$$

$\emptyset$ = normal distribution. Constant K and dispersion S depend on point g.

THE SEQUENCE OF EARTHQUAKES

The frequency-of-occurrence law - $L_g(m)$ - is a function of magnitude and has the following appearance:

$$L_g(m/g) \;=\; \ell_0(g)R(m/g)$$

where

$$\ell g R(m) = \begin{cases} -B(m-m_1) & \underline{m} \leqslant m \leqslant m_1 \\ -B_1(m-m_1) & m_1 \leqslant m \leqslant \bar{m} \end{cases}$$

Parameters of slope B and B_1, value $\underline{m}$ and maximum $\bar{m}$ depend on point g. The parameter $\ell_0(g)$ defines the intensity of flow of earthquakes of magnitude m_1 at point g.

DESCRIPTION OF COMPUTER PROGRAM

The logical design of the computer program whose main task is to compute the distribution function of the total effect, is divided into three main parts:
- Establishing the map of the region;
- Estimation of distribution of the effect from a single earth-
  quake;

- Estimation of the distribution function of the total effect
  applying the formulas shown above.

The first part includes the reading and compact coding of
initial information. The final map of the region is constructed
in the following way. First the geometrical danger zone G and
object zone 0 is read into the computer memory by defining a
number of straight lines constituting the boundary of the different zones which might be of any shape. For each zone specified
all the parameters involved in the analysis and representative
for this region are read into the computer memory. Finally a
rectangular grid is superimposed on the entire area under consideration and all the information pertinent to the model is
now located to all the grid points within this area. This is done
numberically in such a way that the computer decides which zone
the actual grid points belong to (if any at all) and supplies all
the information pertaining to that zone to this grid point and
saves the information in coded form within the computer memory.

Hence for each point of the zones considered one has to
specify the following parameters:
- $O(g)$. A sign (0 or 1) defining whether point g belongs to the
  object 0.
- $H(g)$. A sign (0 or 1) defining as above whether the point
  belongs to the danger zone of an object.
- $\Delta I(g)$. Correction to the intensity of tremors for the actual
  soil conditions.
- $\bar{A}(g)$. Average azimuth of isoseismals.
- $F_{\Delta A}(./e,g)$. Distribution function of the correction to the
  azimuth.
- $L_c(g)$. Intensity (frequency-of-occurrence) of earthquakes for
  some fixed energy e_0.
- $F(e/g)$. Type of distribution function of occurrence of earth-
  quakes as function of energy, i.e., defined by parameters
  $B(g)$, $B_1(g)$, $m_1(g)$ and $\bar{m}(g)$.
- $X_c(g)$. Effect from a tremor of intensity c at point g.
- $P(t/g)$. Dynamics of effect $X_0(g,e)$ as a function of time.
- $I_0(\bar{g}/g,e)$. Average model of isoseismals for normal soil.
- $F_{\Delta Q}(-/g)$. Distribution function of logarithmic correction for
  areas of isoseismic zones.

As might be seen from the presentation above, most of the
information is presented to the computer in terms of random corrections, i.e., with their corresponding distribution function.
Hence if an international cooperation across the different
disciplines such as geology, tectonics, seismology and engineering,
etc., might result in an improved and integrated data base in
such a way that the required parameters can be read into the
computer with some degree of reliability, the present computer
version of the seismic risk model might prove efficient in a

total seismic risk zoning in the North Sea. A similar analysis
has been accomplished for lifelines in the Boston area in the
USA as well as in Italy.

REFERENCES

Caputo, M., V.I. Keilis-Bork, T.L. Kronrod, G.M. Molchan, G. Panza,
     E. Piva, V.M. Podgaetskaya, D. Postpishl. Seismic risk in the
     territory of central Italy, Computational Seismology, Moscow,
     1974.

Keilis-Borok, V.I., T.L. Kronrod and G.M. Molchan. Algorithm for
     the estimation of seismic risk, Computational Seismology,
     Moscow, 1974.

NAME INDEX

Adams, J. 30
Ahorner, L. 10, 88, 98, 101, 110, 111, 306, 386, 387, 390, 396
Aki, K. 310
Allen, C. R. 20, 28
Ambraseys, N. N. 25, 30, 58, 74, 81, 88, 98, 274, 317, 359, 362
Andersen, A. 258
Andersen, K. H. 218
Anderson, D. L. 24, 30
Anon 29
Ansquer 189, 193
Armbruster, J. G. 379, 383
Austegaard, A. 69, 74
Avedik, F. 171

Baker, J. L. 375
Basham 341
Bath, M. 58, 74, 75, 349, 363
Baumann, H. 111
Bazan, E. 30
Beaumont, C. 130
Berger, J. 130
Beringen, F. L. 192, 194
Biot 188, 193
Bjerrum, L. 258
Blume, J. A. 264, 375
Boegner, P. L. E. 29
Bolt, B. A. 20, 28
Bolton 265, 276
Bonilla, M. G. 24, 28, 29, 30
Bonjer, K.-P. 111
Bouma, A. H. 259
Bowers, J. E. 170
Brady, A. G. 288

Browitt, C. W. A. 155, 156, 170, 363
Brown, B. T. 170
Brown, S. F. 218
Buchanan, J. M. 24, 30
Bungum, H. 59, 61, 74, 99, 150, 248
Burton, P. W. 99, 347, 351, 354, 362
Butcher, J. 170
Byrne, D. A. 170

Caputo, M. 404
Carton 189, 193
Caston, V. N. D. 12
Chiarrutini, C. 359, 362
Christie, P. A. F. 160, 164, 170, 362
Chung, D. M. 363
Clough, R. W. 247
Collette, B. J. 310
Cornell, C. A. 355, 362, 367, 375
Costes, D. 99
Cranford, M. D. 170

Davison, C. 57, 73, 349, 362
De Becker, M. 130
De Gijt, J. G. 187, 192, 194
Demars, K. R. 258
Dollar, A. T. J. 57, 73, 337, 338, 349, 362
Doornbos, D. J. 54, 73
Doppert, J. W. Chr. 13
Drake, C. L. 258
Ducarne, B. 130
Durst, H. 111

405

SUBJECT INDEX